Leaf Prints of
American Trees and Shrubs

(*A Modern American Herbal*)

LEAF PRINTS OF AMERICAN TREES AND SHRUBS

(*A Modern American Herbal*)

1974

LITTLEFIELD, ADAMS & CO.

Totowa, New Jersey

Published 1974 by
LITTLEFIELD, ADAMS & CO.
by arrangement with A. S. Barnes & Co., Inc.

Library of Congress Cataloging in Publication Data

Marx, David S.
Leaf Prints of American Trees and Shrubs
(A Littlefield, Adams Quality Paperback No. 279)
Published in 1973 under title: A modern American herbal.
1. Trees—United States. 2. Shrubs—United States. 3. Botany, Economic—United States. 4. Leaf prints. I. Dugdale, Chester B. II. Title.
[QK482.M43 1974] 582'.16'0973 74-3162
ISBN 0-8226-0279-2

Printed in the United States of America

To

Charles Waring Jones,

who carried a birch rod in one hand and a balm in the other while this book was in conception. (D. S. M.)

CONTENTS

THE PLANT FAMILIES

ACKNOWLEDGMENTS

The deepest expressions of gratitude must be reserved for David S. Marx. He had produced all of the leaf-prints, he had developed the basic conception, and he had written a small part of the text before he died. In a sense it is his book. But on the other hand, it was not, at his death, a manuscript. It has taken many months of sorting, assembling, checking, writing new material, and rewriting the old, to the extent that any errors that now remain are my own and I take full responsibility for them.

I wish here to acknowledge the debt owed to my wife, Anne, for doing much of the proofreading and for her understanding when I was "preoccupied" with the work, and to Mrs. Betty F. Storms for her assistance with the typing and indexing.

I also wish to gratefully express my appreciation to Dr. Antoinette M. Anastasia, formerly chairman of the Department of Biological Sciences and now Dean of the Maxwell Becton College of Arts and Sciences of Fairleigh Dickinson University for her constructive criticisms after a very careful reading of the text.

Last, I wish to thank Mrs. Mathilde E. Finch, Associated University Presses' Editor-in-Chief, Scholarly Books, for her editorial skill in changing a rough manuscript into a finished work.

C. B. D.

Introduction

DAVID SAMUEL MARX (1907–1968)

Visit any generalized farm and you will return with a deeper respect for the plant kingdom. Sit in a wooden chair. Go to your cupboard and open a can of peas. Read your newspaper. Use a plastic dispenser bottle, mucilage, cellophane tape—*you* finish the list, if it ever can be finished!

We use plants for most of the necessities of life —foods and drugs, clothing and construction. Sometimes we get smarter than nature and synthesize such things. What from? From other plant materials. It has been truly said that we are merely guests on this planet—guests of the plant kingdom and a thin layer of topsoil!

David S. Marx labored with love for many years over the manuscript of a book that he proposed to call *An Herbal for Americans*. It would tell Americans something about what the plant kingdom was doing for them. Hopefully they would better understand some of the interrelationships in the web of life. Before his task was accomplished, however, David Marx died. The manuscript remained in the publisher's storage room for several years until, in the spring of 1971, a decision was made to rework the material and to go on with its publication.

Marx was born in Troy, Ohio, in 1907. He attended the University of Cincinnati and the University of Chicago. However, in the field of botany he claimed to be self-educated. It was evident from reading the material he had prepared for the *Herbal* that over fifty years spent in botanic gardens, in studying herbarium specimens, in reading, and in doing much practical nursery and horticultural work, had, indeed, given him a superb botanical education.

During his varied career Marx had been a nature counsellor in summer camps, conducted a series of nature talks on radio, wrote a newspaper column, and published a dozen magazine articles. One of these appeared in the September, 1954, issue of *School Arts* and was entitled, "Making Prints from Leaves." It was with the highly refined techniques described in this article that the leaf-prints reproduced in this work were originally made. The stored box of material described above included hundreds upon hundreds of exquisite leaf-prints. Tree leaves, fern fronds, mosses, grasses, flowers—every sort of botanical material was recorded by Marx's leaf-print technique.

LEAF-PRINTS

If you should wish to make some of your own leaf-prints, it would be well to read David Marx's directions in the article mentioned. But the technique is relatively simple and here is the essence of it. The leaf should be fresh and quite flat. If it must first be brought home from the field, keep it flat in a plant press or large catalogue-type book. An oil-base linoleum-block printing ink is applied very sparingly to a piece of glass of convenient size, and is then spread evenly over the surface of the glass with a rubber roller (brayer). Apparently one of the critical elements of the technique is the interpretation of the word "sparingly" used above. Trial and error must be the teacher, but the error is quite likely to be in the use of too much ink rather than too little. Now lay the leaf on the inked glass and carefully run over it with the roller. Turn the leaf with a forceps and do the other side. Next place the leaf between two sheets of a high-gloss, magazine type of paper and feed the sandwich carefully between the wringing rollers of a washing machine. In these days of the spin-dryer, locating a roller-type wringer may well be the most difficult part of the whole process. A hand-turned wringer from a second-hand shop could be the answer, although this type may require a "third hand" to crank the handle as the sandwich is being fed in. Careful separation of the two sheets of paper and the removal of the leaf with a pair of forceps completes the process. You now have an exact replica of both sides of the leaf. The finest veinlets, bristles, and even hairs should be visible—unless you used too much ink!

HERBALS

The old European herbals were gropings in the direction of understanding the uses and classification of plants. For instance, Agnes Arber in her thorough history of the development of herbals during the thirteenth to fifteenth centuries quotes one Nicholas Culpeper as writing "Mushrooms are under the dominion of Saturn and if any are poisoned by eating them, Wormwood as an herb of Mars cures him, because Mars is exalted in Capricorn, the House of Saturn; and this is done by sympathy."[1] In those days it was more important to know what a plant was "good for" than how many stamens it had. Plant names like Dog-bane and Louse-wort had their origin in this point of view. Thus, this orientation toward uses and values gave these old herbals (and hence also the modern ones) a very anthropocentric and teleological cast. When Mr. Marx suggests, therefore, that red berries are "good for" attracting birds to a particular tree which, in turn, is "good for" the dissemination of its seeds, he is using "herbal language." The preference of modern scientists to say that such phenomena have come into existence or, at least, remain in existence because they have "adaptive value" does not materially change the facts or improve our understanding of them.

After the age of "good for," came the age, in the eighteenth and nineteenth centuries, of the development of abstract botany, upon which the sciences of agriculture, forestry, and even much manufacturing rest. No one can use a plant intelligently and completely until he knows all there is to know about it. This book is a swing back toward utilization rather than the theory of plants. We can make a better herbal now than in the earlier centuries because we have a better framework upon which to hang the details about food, drug, industrial, and other uses.

To indicate the framework upon which this *Herbal* is built, Marx customarily drew an illustrative geometrical figure. It consisted of a square block, which was the *Herbal,* resting upon four solid foundation-stones. These bore the identifying names of Bailey, Fernald, Hill, and Benson. The complete titles of the four great works to which he referred will be found in the bibliography. This *Herbal* indeed rests heavily upon these foundation-stones. The sequencing of the families, for instance, as well as the estimates of the numbers of genera and species that are included in the families, is largely as they are proposed by Bailey. Much of the information concerning the commercial uses of plants was from Marx's own experience, but some of it is from Hill, and the rest from more modern sources. From Fernald comes information about plants other than those in cultivation, and from Benson much about the structure of flowers, fruits, and seeds, and the newer ideas in taxonomic relationships.

1. *Herbals, Their Origin and Evolution* (Cambridge, 1953), p. 263.

TAXONOMY

In some of the botanies of the preceding century, what we now call families were called "natural orders." Watch out for this if you are perusing the old texts.

This herbal is concerned mostly with families, genera, and species. These are groupings of plants so defined that each larger taxon includes in its definition all the characteristics found in the smaller taxa within it. The Oak genus (*Quercus*), for instance, includes in its definition every characteristic exhibited by the various Oak species (*Q. nigra, Q. alba,* etc.). The Beech-Oak family (FAGACEAE) is comprised of the Beech genus (*Fagus*), the Chestnut genus (*Castanea*), the Oak genus, and three others. The definition of the Beech-Oak family, therefore, must include all the characteristics seen in the six genera. There are some 600 species in these genera and it is the characteristics that they share in common that link them into a family.

A family can consist of a great many genera or it may contain only one genus with just a single species. Such is the case with *Ginkgo biloba,* which is the only living representative of the family GINKGOACEAE. In any case, a family is a taxon containing whatever genera an authority believes belong in it. To some extent his decisions are based upon "natural" relationships; to some extent they depend upon his judgment. Furthermore, there are differences of interpretation of these "natural" relationships from one authority to another.

The layman might well be sceptical of the exactitude of Science (with a capital "S") when he reads that there are about 4,500 to 5,000 species of plants in the Madder family (RUBIACEAE).

In a society where the automotive industry can tell us precisely how many cars were produced last year and probably, if required, exactly how many nuts and bolts they contained, it seems rather inadequate of the taxonomists that they can only ascertain the number of species in this family to within $\pm$ 250. After all, can't they count?

The difficulty lies in the nature of the concept "species." It is difficult to count them because it is difficult even to define one to everyone's satisfaction. A twofold definition is rather widely favored. It starts out by stating that all the plants (or animals) *of any one kind* growing in a particular location on a fairly stable basis and preferentially interbreeding with one another constitute a *population.* (All the Ponderosa pine on the Cartwright ranch, for instance, form a population.) The sum of all the populations, *of any one kind,* on earth is a *species.* The members of the species also preferentially interbreed with one another, or at least they do if it is practicable. Geographical separation, for instance, can make completely random interbreeding within the species impracticable.

One will recognize immediately that the weakness in this definition lies in what is meant by *of any one kind.* Another characteristic of our society is the feeling of its members that practically all things occur as discrete units that can be counted, sorted, filed, retrieved, and disposed of. If it were not so, how could any complex "Parts Department" be possible? Many people have been philosophically persuaded that since plants and animals were also created *after their own kind,* they, too, should be amenable to counting, sorting, filing, and so on. But a study of the life sciences is quite apt to convince us that all life has been evolving from preexisting life for millions of years by way of a long, on-rolling series of small changes. Hence, as the evolution of the Beech-Oak family proceeded, there was no reason why it should, and, indeed, no possible way that it could, exactly parallel the evolution of the Maple family, for instance. Neither the genera nor the species in the two families can be *exactly* the same natural entities. In other words, when we say that there are 20,000 species in the Orchid family and 200 species in the Pine family, we are not at all certain that the 20,000 and the 200 are the same *kind* of units. And when we consider the relationships between hawks and bacteria, we are quite certain that they are not the same!

And so we return to our original point about the subjectivity of taxonomy. The degree of genetic relationship between plants, which is what the word species should stand for, is actual, but often difficult to measure; but the way individual botanists evaluate the known facts is personal and sometimes arbitrary. Taxonomists who tend to group somewhat heterogeneous plants into one taxon are called "lumpers," while those who prefer smaller, separate taxa are called "splitters." Of course, a particular botanist may be one in one instance, and the other in another.

Fernald lumps Cypresses and Junipers into the Pine family and Mountain-ashes, Apples, and Choke-cherries into the Pear genus, while Bailey separates them. With notable exceptions, of the four basic authorities cited in this *Herbal,* Hill and Fernald are "lumpers" (i.e., conservative) and Benson and Bailey are "splitters" (not radical, but liberal).

THE PLANT LIFE CYCLE

Most of us are familiar with a reproductive cycle in which sexual parents, male and female, produce babies, also male and female, who in due course mature and repeat the cycle. In most animal species the offspring bear enough physical resemblance to their parents to indicate the species to which they belong; the kitten is obviously a cat.

Those who have had no botanical training may be surprised to learn that the life cycle of the plants described in this book is quite different from that found in most animals. The plant life cycle can be illustrated by this bizarre example: A pair of sexual parents produce an asexual "baby," whose appearance is as different from themselves as a floorlamp might be! But let us leave bizarre imaginings immediately and consider the botanical actualities.

In the life cycle of higher plants, a pair of sexual parents consisting of a female (*Megagametophyte*) that produces one egg and a male (*Microgametophyte*) that bears two sperm nuclei, both of which parents fulfill functions similar to those of their counterparts in the animal kingdom, does occur. However, both of these sexual parents are very minute and are best seen with the aid of a microscope. The male plants (*Microgametophytes*) are commonly called *pollen grains.* The female plants are tiny masses of cytoplasm containing eight nuclei that are found within little *ovules* in the *ovary* of the flower. (Reference can be made to the illustration of the structure of the flower that follows.) The fertilization of the female's egg nucleus, which is one of the eight nuclei just mentioned, is accomplished by one of the sperm nuclei of the pollen grain. This results, as usual, in a pregnancy, the development of an embryo, and eventually a "baby." When we first see this embryo it will be as part of a seed, although we may also buy it bottled as "Wheat Germ." When the embryo in its seed reaches the ground, it will continue its growth through the seedling stage to the mature shrub or tree.

But the nature of this "baby" that develops from the fertilized egg and finally grows in our garden is quite different from its counterpart in the animal kingdom. It is sexless, and it is distinctly different in appearance from either of its parents. They were sexual and minute; this offspring is asexual and perhaps as large as a "mighty oak." Because this asexual offspring when mature produces *spores,* not egg or sperm, it is termed a *Sporophyte*. Spores are cells that do not need to be fertilized in order to develop. Develop into what? Into pollen grains and megagametophytes, the sexual parents of the next sporophyte generation.

A complication. There must of necessity be two kinds of spores produced by the sporophyte: One that develops into a pollen grain, the other into a female gametophyte. In many plants the two kinds of spores (called *megaspores* and *microspores*) are produced in the body of the same individual. However, in other species, the production of the two kinds of spores has been separated and they are produced on different plant bodies called, respectively, *Megasporophytes* and *Microsporophytes.*

In technical terms, the plant life cycle is often expressed by this circular pattern:

This summary of the life cycle will be repeated in a different form at the close of the section on Fruits and Seeds.

In terms of our practical experience, this discussion of the reproductive life of plants indicates that trees and shrubs as we see them and plant them are part of the sporophyte phase of their life cycle and they are sexless. The terms "ovule" and "ovary," then, are misnomers because they are words with female implications, while the plants that bear them are sexless. Furthermore, when we buy certain trees (e.g., Holly), the nurseryman will urge that we acquire both a "male" and a "female" tree to assure a better crop of berries. Why does he use this terminology for the sporophyte generation? Either because he is not familiar with the story told above, or he doesn't believe that you are! In any case, it is too cumbersome to advertise

MEGASPOROPHYTE AND
MICROSPOROPHYTE
HOLLY TREES

FOR SALE!

So be it. In this *Herbal,* we will call them male and female, too.

FLOWERS

The plant embryo that we have been discussing is essentially a small, solid, bar of cells. The cells in the central portion of the bar elongate and so increase its length. The cells at either end of the

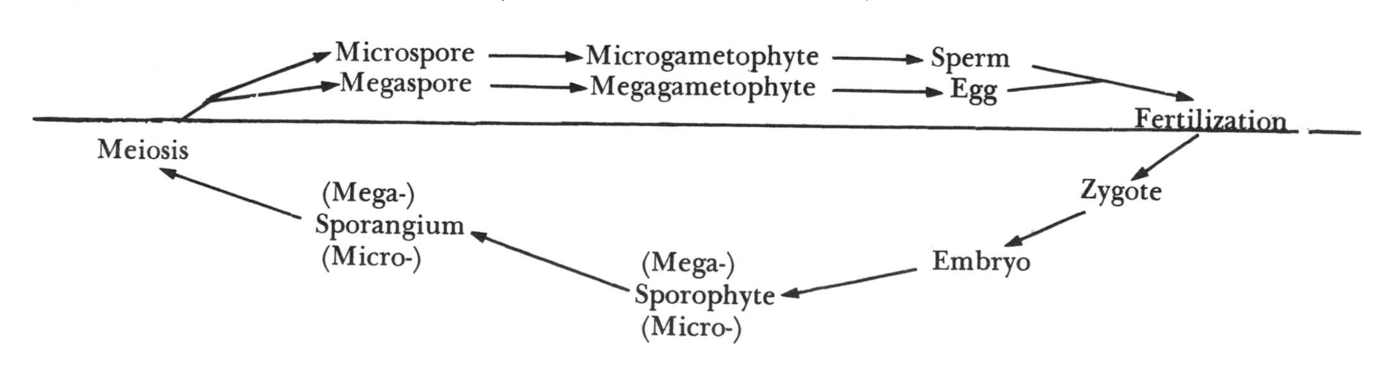

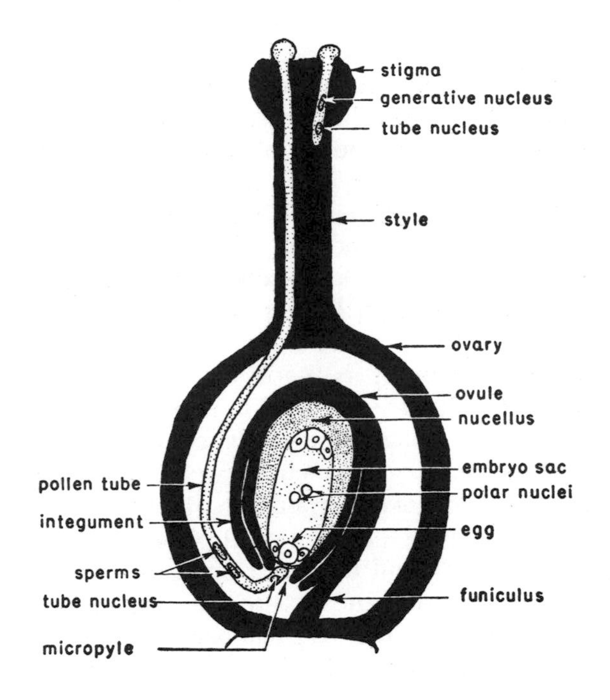

bar continuously multiply and produce new cells. Of these new cells, the more central ones also elongate and so the plant grows taller. The lower end of the bar gives rise to roots, the upper end to stems and leaves. The two ends of the bar are called *meristems.*

During the growing season the stem meristem produces a), more of itself, b) leaves, and c) offshoots of itself, which are the side branches, in the angles between the leaves and the main stem. This is called vegetative growth. There comes a moment, however, when this kind of growth stops and reproductive growth is substituted. A flower is initiated. The triggering mechanisms that are able to bring about this change in the type of growth are fairly well known and having been described in most general biology and botany texts will not be discussed here. The point to be noted, however, is that a flower is a modified branch and the flower parts are generally accepted as being modified leaves.

Perfect flowers consist of four kinds of such modified leaves. The sequence followed in the listing below starts with the lowest (or outermost) modified leaves involved in the flower's structure and proceeds upward (inward).

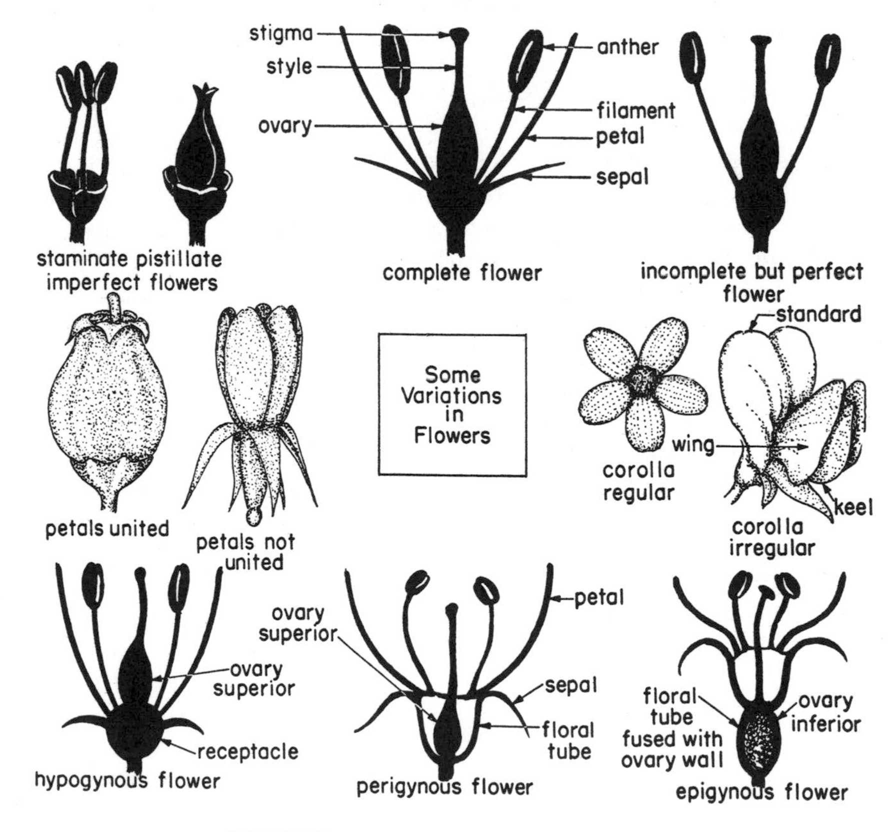

* Fig. 19-20 (p. 262) from Henry T. Northen, *Introductory Plant Science* (New York: The Ronald Press, 1958). Reprinted by permission of the author and The Ronald Press Company.

** Fig. 20-8 (p. 276) from Henry T. Northen, *Introductory Plant Science* (New York: The Ronald Press, 1958)). Reprinted by permission of the author and The Ronald Press Company.

1. The *sepals,* being outermost, cover the developing flower bud. As seen on a Rose, they are green, but on a Day-lily, yellow, so their color varies. When colored, they may be as attractive as the petals. Collectively, the sepals are called the *calyx* and they are often more leaf-like in construction than any of the remaining flower parts.

2. The *petals* are either white or colored. Their value to the plant appears to depend upon their ability to attract the insects or birds that effect pollination. From our point of view they give the flower most of its beauty. The *corolla* denotes all of the petals collectively. The term *perianth* (or *floral envelope*) is used to describe both the sepals and the petals, especially in those flowers where they gradually intergrade or are identical in appearance.

3. The *stamens* (collectively, *androecium*) consist of *anthers* and *filaments* or stalks. Within the anther are the sacs (*microsporangia*) that produce the *microspores* discussed in the previous section. These are the spores that develop into *Microgametophytes,* which are the male plants or *pollen.*

4. The innermost (or topmost) modified leaf or leaves are called *carpels* (collectively: *gynoecium*). The structure formed from one carpel, as in the pea plant, or from three carpels (lily), is the *pistil.* The pistil, in turn, has three parts: the *ovary* at its base, a tubular *style* above, and a *stigma* at the upper end of the style. Within the ovary are one or more *megasporangia,* each of which produces a *megaspore* that grows into the minute female plant (Megagametophyte). The Megagametophyte was discussed at length in the previous section on life cycles and it will be remembered that this is the female plant that produces the egg, which is fertilized by the sperm nucleus from the pollen grain. The fertilization of the egg results in a *Sporophyte* embryo and soon thereafter a *seed.*

The portion of the stem from which these four kinds of modified leaves arise is called the *receptacle;* that portion of the stem immediately below the receptacle is called the *pedicel.* Another set of modified leaves (*bracts*) occurs at the node below the pedicel. A *node* is a place on a stem where leaves can arise. Continuing down the stem, the next section below the bracts is the *peduncle.*

FRUITS AND SEEDS

Fruits of many kinds are produced by trees and shrubs. This is true because the appearance of the mature fruit rests upon a number of varying factors. Three of these will be discussed briefly.

A fruit can be defined as a ripened ovary and (on occasion) its associated parts. This means that the fruit, as we know it, *may* consist of tissue that in the flower was identified as being the ovary, the receptacle, the floral tube, and so on.

Second, an ovary is made up of one or more carpels (modified leaves). To visualize a carpel in cross-section, imagine a short curved line that begins and ends with a small circle. These circles represent the megasporangia, which are found along both edges of the carpel. Open a pea pod along its suture and note that one seed is attached to the left edge and the next to the right-hand edge of the carpel. If the ovary has been built up from more than one carpel, for instance three, the component carpels can be assembled in several ways. They may be joined together edge to edge to make one closed circle. A slice of cucumber illustrates this very clearly. Another possibility is that each carpel be individually rounded and sealed and then they can all be united into one body. Citrus fruits illustrate this arrangement. Hence, the basic structure of the ovary affects the final appearance of the fruit.

The last influencing factor to be mentioned is the physical or cellular construction of the carpel itself. The carpel, as seen in the ovary wall, usually exhibits two more-or-less distinguishable layers—an outer *ectocarp,* and an inner *endocarp;* together, especially when indistinguishable, the *pericarp.* These two layers may have a variety of cellular textures such as fleshy, dry, leathery, woody, stony, and the like. These textures, alone or in combination, determine the basic nature and edibility of the fruit.

In summary then, a fruit is derived from one or more ovaries and may include other associated flower and stem parts. The ovary itself, as well as the other contributing parts, may have one or several of a variety of textures. Thus from the flower a fruit is developed.

Below, in outline form, is a discussion of some of the more common types of botanical fruits. Not all of the types shown occur on the trees and shrubs included in this herbal, but, on the other hand, this is not a complete listing either.

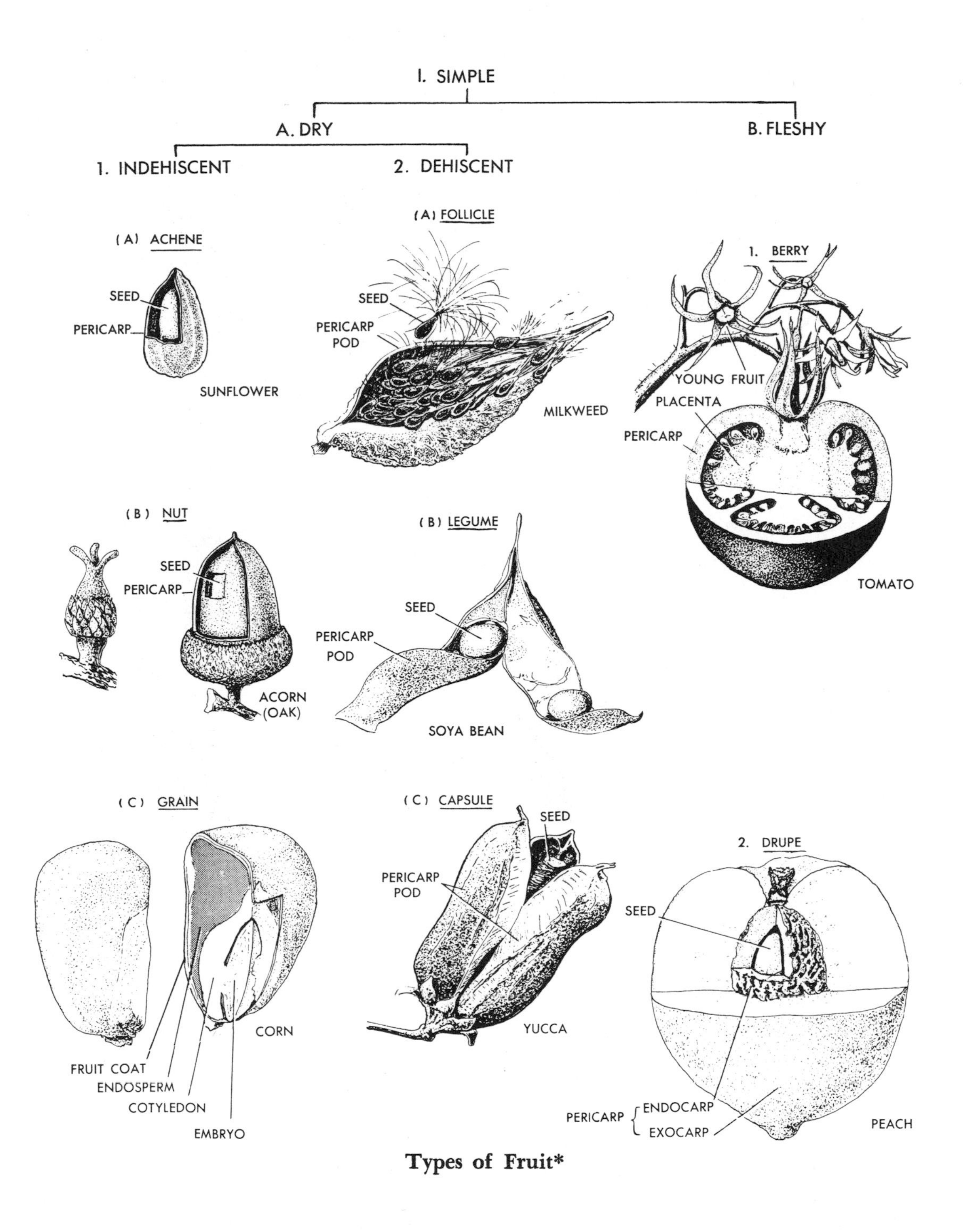

Types of Fruit*

(Illustration continued on following page)

* Fig. 25.11 (pp. 480-81) from *Biology,* third edition, by Willis H. Johnson, Richard A. Laubengayer, Louis E. DeLanney and Thomas A. Cole. Copyright © 1966 by Holt, Rinehart and Winston, Inc. Copyright © 1956 and 1961 by Holt, Rinehart and Winston, Inc. under the title *General Biology.* Reprinted by permission of Holt, Rinehart and Winston, Inc.

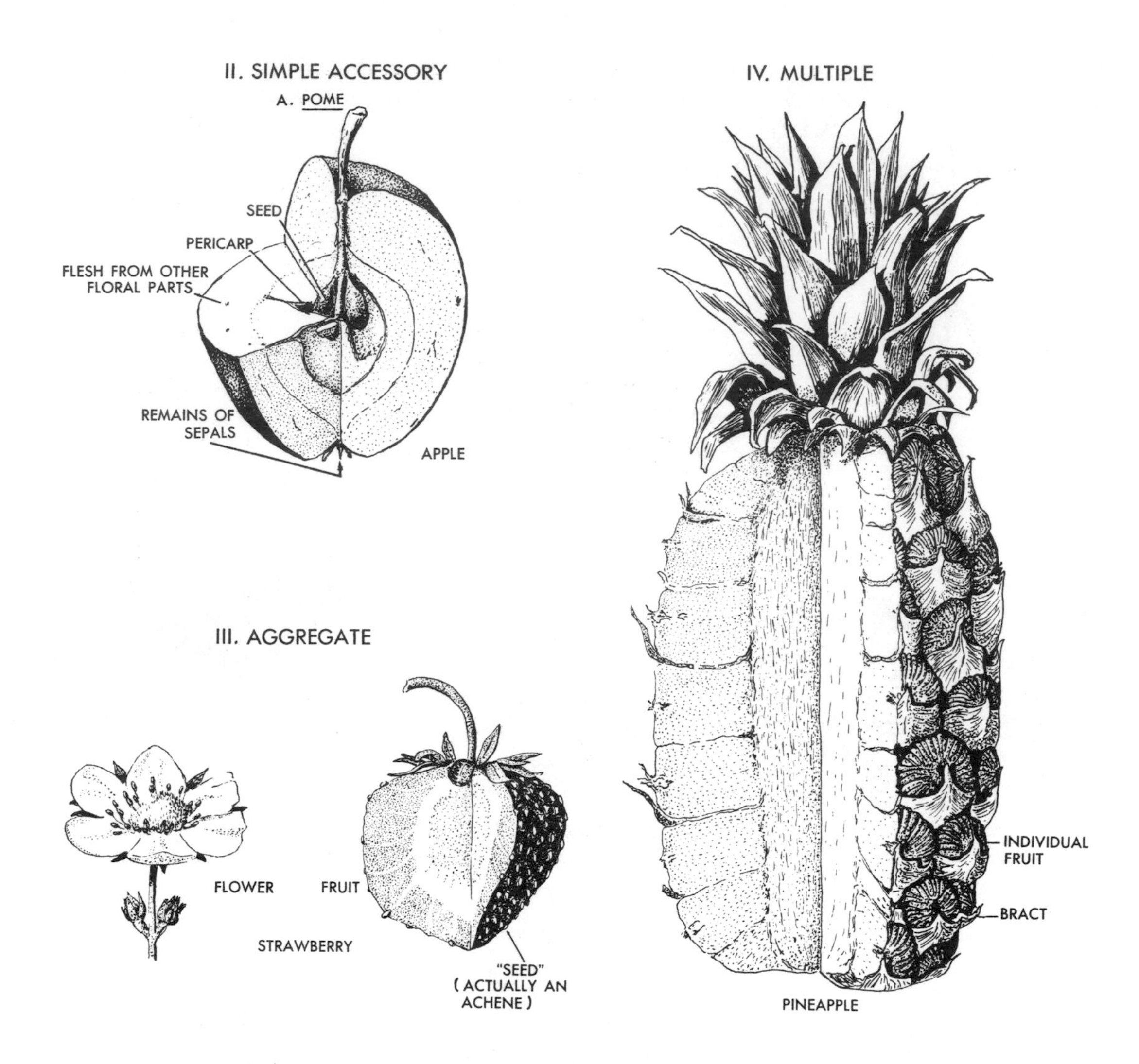

A. Dry Fruits
  1. One-seeded and indehiscent (not normally self-opening along a natural line).
    a. *Achenes*—The seed is tightly inclosed in the pericarp (sunflower).
    b. *Nuts*—Similar to achene and usually larger and with a hard, usually thick, "shell" (pecan).
    c. *Samaras*—Have a long thin flat wing (maple "nose").
  2. Several to many-seeded and dehiscent
    a. From a single carpel
      1. *Follicles*—Dehisce only along one line: Either along the suture where the edges of the carpel meet (milkweed) or along the midrib of the carpel (magnolia).
      2. *Legumes*—Dehisce along both the suture and the midrib (pea).
    b. From more than one carpel
      1. Capsules—Lengthwise dehiscence along regular lines thus separating capsule into valves (iris).

B. Fleshy Fruits
  1. *Berries*—One to many seeds inclosed in a fleshy endocarp. Skin is generally thin (tomato) but may be somewhat thicker (avocado).
     Special kinds of berries are:
    a. *Pepos*—Many seeds inclosed in a fleshy endocarp with a hard or leathery rind (watermelon).
    b. *Pomes*—Several seeds which are inclosed within the ovary wall (core). Edible portion is derived from the floral cup (apple).
  2. *Drupes*—The stone fruits. Usually one seed. The exocarp is fleshy and the endocarp stony (peach).

Note: Some fruits that are commonly called "berries" (strawberry, raspberry, and blackberry) are not botanically true berries. The raspberry and blackberry are *aggregate fruits* formed from a group of ovaries produced in a single flower. In the strawberry, the botanical fruits are the achenes embedded in the flesh and would be described by the layman as the "seeds." Part of the fleshy substance is derived from the "associated parts" of the flower.

The flowers of raspberries and blackberries have many spirally arranged pistils on the receptacle. Each develops into a drupelet. In the raspberry the mass of drupelets pulls away from the receptacle when the "berry" is picked; in the blackberry, it does not.

THE LIFE CYCLE (REVISITED)

Having discussed the structure of the flower and the types of fruit, we should once more examine the life cycle of the flowering plant, building into it more information than in the first version. In this rendition:

a. Male gametophyte structures and relationships are printed in ordinary type-face.
b. *Female gametophyte structures and relationships are italicized.*
c. **The sporophyte generation is in bold-face type.**

a. Microspore **(formed in microsporangium of anther)** develops into pollen or microgametophyte; pollen tube grows from pollen grain and bears two sperm nuclei.

b. ***Megaspore*** **(formed within ovule in ovary)** ***develops into tiny gametophyte; one of the eight nuclei in her body is the egg.***

c. ***Egg*** (fertilized by sperm) **develops into embryo; embryo is contained within seed (now a "ripened" ovule) which is enclosed in the fruit (now a "ripened" ovary). Seed is dispersed and then continues its growth on the ground and becomes the sporophyte tree with flowers. Anthers of the flower produce . . . .** Microspores. **Ovules of the flower produce . . . .** ***Megaspores.***

SEEDS AND SEED DISPERSAL

In the sections "The Plant Life Cycle" and "The Life Cycle (Revisited)," we discussed the seeds as an early stage in the development of the sporophyte generation. Plants vary in just how far the fertilized egg develops before the seed comes to maturity within the fruit and rests. One may sometimes find a pea seed within its pod bearing a fairly well-developed root. That seed has progressed beyond the embryo stage and is already a "baby" or seedling plant. If this seed were to continue its growth within the pod, a very odd situation would shortly exist. A young pea plant would be growing out of a pod atop an old pea plant. Or perhaps a young pine tree could be seen growing out of a cone atop an old pine tree. This leads us to an appreciation of the function of the seed. It is the place in the life cycle where the sporophyte of the next generation returns to the ground. A seed could accomplish this by merely plopping straight downward to the soil. And then what would we have? A young pea plant or a young pine tree growing directly *beneath* its forebear. That is almost (but not quite) as unsatisfactory as growing atop its parent. So, finally, the function of the seed is to get the next sporophyte individual into the soil at some distance from the plant that produced it. Thus we use the terms "seed dispersal" and "seed dissemination" to cover this concept. A study of the methods that have evolved for seed dispersal is a fascinating one. Only some of the possibilities can be suggested here.

It will be recognized, of course, that ofttimes the term *seed dispersal* actually refers to fruit dispersal. The squirrel may bury the whole fruit of the nut tree and the bird may eat the whole berry. The seed dispersal is then incidental to the fate of the fruit. In what follows, the term *seed dispersal* will be used in the wider sense.

In most of the upland Pines and Spruces, the small, winged, "helicoptering" seeds are carried by the wind. However, in a few Pines, larger seeds are transported by man, ants, squirrels, and other animals. In the Bald-Cypresses, which grow in water, the cones do not open but float, still bearing their seeds, into new localities. In Junipers, the cones are modified into succulent berry- or drupe-like cones distributed by birds and other animals. It may well be possible that the rank flesh of the fruit of the Gingko was once relished by some prehistoric coprophage animal that thus distributed the seeds.

Rather than individual examples listed *ad infinitum,* the general types of seed-dispersal can be summarized with these four categories:

1. By animals via burying; by ingesting the fruit but not digesting the seeds; by catching on fur; by sticking to mud on feet, etc.
2. By floating in water
3. By air via helicopter-type winged seeds; by parachute-type devices as in Milkweed; and by the sheer force of wind on fine seeds as in Orchids
4. By mechanical means as in the sudden dehiscence of pods that catapult the seeds into the air

Beyond the mere fact of the physical dispersal of seeds that can be objectively described, there is another interesting area, which is, however, conjectural or at least less amenable to objective study. It is, moreover, an area in which David Marx was greatly interested and had given considerable thought. The basis of the following discussion has been gleaned from his notes.

It is quite obvious that the pulp of the berry nourishes the bird and that after the pulp has been digested away, the unharmed seeds that have passed through the alimentary tract are deposited on the ground where they may germinate. Indeed, some seeds will not germinate unless they have made this passage and seedsmen must simulate the abrasive action of the digestive system on the seedcoat by mechanical manipulation. Likewise, the walnut tree supplies the squirrel with nuts for his winter fare, but in the act of burying them the squirrel is effectively planting the next generation of trees because he does not dig up all that he buried. It is quite within reason to claim that the bird and the squirrel possess either learned or inborn behavior patterns that lead them to the berries and the nuts. But what about the trees and shrubs? Can they be said to be actively engaged in enticing the animals to their fruits to the end that at least a portion of the seeds will fall on fertile ground? It need not be so teleologically expressed. One can ask: Do plants have adaptations evolved through natural selection whose survival value is that they insure the participation of animals in the seed-dispersal process?

Marx was quite certain that such adaptations do exist. It was his opinion, for instance, that in the fall the red leaves signal birds from afar to plants that produce berries and that the yellow leaves signal quadrupeds to the nuts. Furthermore, a considerable number of the northern Oaks, the native Beech, the Sugar Maple, and at least some of the Hickories carry the lower leaves through the winter. In many an old-time romantic novel there are passages describing the soughing of the wind through the leaves. But these lower leaves *could* be ripened and abscissed by these trees just as easily as the upper ones, so perhaps their retention may have adaptive value. Could these marcescent leaves perhaps be signaling squirrels to their fruits?

CALENDRATION

We can take a further step into conjecture. Is there any evidence for adaptive value in the timing or calendration of the stages of development of plants?

There are two calendrations in our flora. One has to do with the stringing out of the periods of flowering so that bees will have their food and flowers their pollination over the whole season. The other, equally as important and almost as obvious, has to do with the overlapping procession of fruits, berries, nuts, and other seeds that assures a constant food supply to the animals and better seed dissemination for the plants. Note, for example, the Strawberry, Raspberry, Blackberry sequence. Another interesting series involves the nut trees. Even before the other tree leaves fall, perhaps as early as August, Buckeyes ripen and fall (these being considered nuts for the moment) . Buckeyes are extremely acrid when fresh and most of them are buried by gray squirrels after the first tentative nibbles. Black walnuts come next and delay their destruction with an acrid pulp that the squirrels prefer not to broach. But, once the pulp is off, the shell of the nut exhibits fine intaglio striations that give the rodents a better grip. The last in the series is the Hickories, of which, if we consider just the Shagbark, we see the sharp upward-and-downward-pointing shards of bark discouraging squirrels from climbing the tree to molest unripe nuts. The nuts themselves are protected by hard hulls. By the time the hulls dehisce and the nuts are mature, the Buckeyes and the Walnuts in the area have been buried or consumed.

But perhaps this is only conjecture. In any case a vast and largely unexplored domain exists wherein very few objective studies have as yet been made.

TREE BARK

The Redwoods and their kin have been subject to forest fires for millions of years. In the course of evolution it has become standard practice for these, and many other trees to develop bark of high insulating value, relatively fireproof, as a protection against this constant and powerful enemy.

The Oaks are also among the trees rather resistant to fire damage. Two species stand out in this respect and two great naturalists independently told their story. In South Florida Simpson tells of the salamander properties of the Live Oak in the fire-regulated struggle between the hammocks and the pinelands. In Wisconsin Leopold explains how the corky bark (even on the twigs) enables the Bur Oak to stand up against prairie fires as outposts of the forest.

The bark of the Cajeput tree (in the Myrtle family) is surely among the most efficient fenders of groundfire; all indications are that this tree so developed because of the prevalence of fire in its native Australia. The common Elm of the Eastern States has corky bark. The Winged Elm of the South has a remarkable webbing of cork in the crotches of its twigs. These may also be fire-fending devices. Surely this would make a worthwhile study project.

If one were to ask, "Why is Birch bark white?", he would again be asking a question whose answer would be conjectural and probably unprovable. Note that when the snow is on the ground and the white trunks of Birches are hazed by their dark twigs, these trees are very hard to see in the distance, and especially so by poor-eyed, bark-eating deer and moose. Why could not the white bark be a matter of plant camouflage as valid as the examples of protective coloration so often seen among the animals? An extension of this explanation comes in the late spring and early summer when the white bark is quite conspicuous against the greenery—but the cervines are not eating bark but the more succulent leaves by then.

IN CONCLUSION

This introductory chapter has been an attempt to present to the reader some established facts and some unestablished possibilities from the field of general botany so that he can better utilize this book in the development of his interest in our useful trees and shrubs.

Someone has defined education as a liberation from the prison of the self. We have an inclination to be full of our own problems, to worry about our own identity, to seek to improve our own condition. In coming to know and understand these other members of the living scene, we are helping to crumble the walls of our prison.

Leaf Prints of
American Trees and Shrubs

(A Modern American Herbal)

The Plant Families

FAMILY 1

CYCADACEAE

THE CYCAD FAMILY

The CYCADS apparently first made their appearance in the Permian period. In subsequent geological ages they were much more widely represented than at present. Today there are only some 9 genera and 85 species of these palm-like plants still existing. Oddly enough, the CYCADS have nothing in common with the Palms except their general appearance, and they are not related to the Ferns even though their leaves unroll circinnately.

1.1

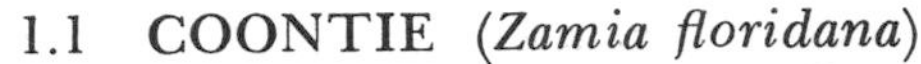

1.1 COONTIE (*Zamia floridana*)

This ZAMIA is a native Florida CYCAD of much historical and some commercial significance. In the North it is an attractive ornamental that requires indoor culture. The stem contains an abundant supply of a substance called Florida arrowroot. It is one of the more easily digestible starches but before it can be utilized a poisonous alkaloid must be thoroughly washed out. The plant was once made to yield a sizable crop. Since it took so many years for the usable part of the plant to reach maturity, the industry was eventually abandoned.

1.2 SOUTH FLORIDA CYCAD
(*Cycas circinalis* L.)

There is a splendid collection of CYCADS available for study in the Fairchild Tropical Gardens on the outskirts of Miami. In this species the seeds are poisonous.

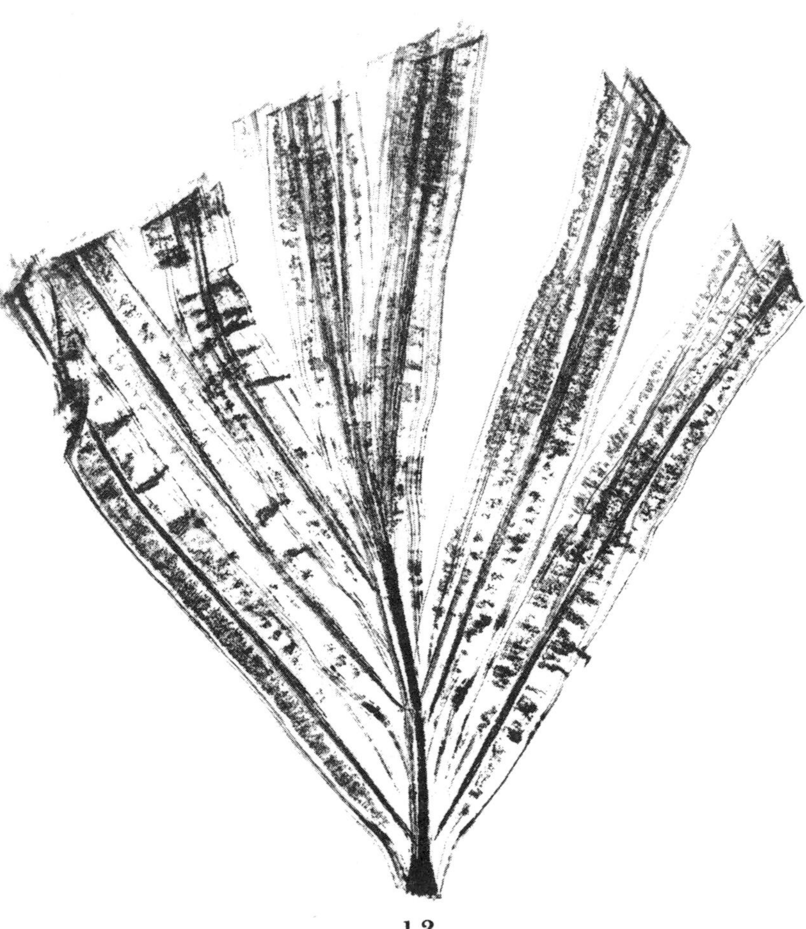

1.2

GINKGOACEAE

The Ginkgo family consists of one deciduous, broad-leaved, and resinous tree that is native in China but now widely cultivated elsewhere. It is believed that it no longer grows wild in its home range even though occasionally travelers report having seen it there. In that sense it can be considered extinct and it is often referred to as a "living fossil." An interesting feature of this tree is its retention of the use of swimming sperm cells to fertilize the eggs rather than the utilization of sperm nuclei carried in pollen tubes. In its original native forests where the male and female trees were in contiguity, the spermata were carried from one tree to the other during periods of rain. As a matter of observation, however, one can note curb-side plantings of female trees, bearing viable seeds, with no males in evidence.

2.1

2.1 GINKGO (*Ginkgo biloba* L.)

The wood of the Ginkgo tree is extremely soft and easy to saw. Its bark is also soft and patterned into a lovely design worthy of being used creatively in designing plaques. The tree is hardy and is practically pest-resistant. The fruit pulp has a strong disagreeable odor, but the seeds within are sweet, white, and edible. Some people may be allergic to them, however.

This family of evergreen trees and shrubs consists of approximately 15 species combined into 3 genera. The leaves are needle- or scale-like. The hard seeds are enclosed by a soft, sweet, scarlet-colored pulp that is an outgrowth of the stalk of the seed. This outgrowth is called an aril.

THE YEW GENUS (*Taxus* L.)

In the genus TAXUS the pulp (aril) does not completely enclose the seed. The species of YEWS are so similar to one another that some authorities believe there is only one species with geographic variations. The male and female flowers are on separate plants and both are needed before berries will set. In practice, since the pollen appears to travel on the wind, the male may be on a neighbor's property.

The foliage and other green parts of all YEWS are very poisonous to livestock. The pulp of the seed is very sweet and quite edible, but the seed within is extremely poisonous and must not be swallowed with the pulp. Yew roots and bark will produce a red or purple dye, according to the mordant used.

3.1

3.1 JAPANESE YEW (*Taxus cuspidata* Sieb. & Zucc.)

This is the YEW almost universally sold by the nurseries in the northern states. As a tree it may attain a height of 30 feet or more. Usually it is used as a shrub for shady exposures. There is no standardization in height and both short and tall forms are available.

ENGLISH YEW (*Taxus baccata* L.)

This species can be distingushed from the one above by its softer and more crowded foliage. The leaves are not arranged stiffly in one plane (distichous) and their undersides are yellowish rather than a true green.

The English Yew is particularly famous as the bow-wood of the English yeoman. Because of the poisonous nature of its foliage, Yews were planted only in cemeteries and other enclosures. This protected the livestock, which in those days often ran loose.

AMERICAN YEW (*Taxus canadensis* Marsh.)

The foliage on this Yew is smaller and glossier than either the English or Japanese species. This native Yew has been observed in a swamp near Sidney, Ohio, where it formed a ground cover, and on limestone rocks south of Rochelle, Illinois, where it formed distinct bushes. One might have taken these to be separate species.

WESTERN YEW (*Taxus brevifolia* Nutt.)

This species, found in the western U.S., is capable of growing fifty feet high or more. However, there is also a dwarf bush by the same common name (and possibly the same species) being sold by midwestern nurseries.

HATFIELD, BROWN, and HIX are other cultivated varieties. They are hybrids that are propagated vegetatively and should come true to form. Beware of anyone who says they will stop growing at some specific height. Most of them will top ten feet in good soil. If you are buying the SPREADING JAPANESE YEW, it would be wise to see the stock specimen; then you will know what to expect.

The PODOCARPS are mostly native in the warmer regions of the southern hemisphere. The family consists of 7 genera and about 100 species. All are evergreen shrubs and trees that bear needle- or scale-like leaves. The general similarity of the Yews and the Podocarps is indicated by the fact that at one time some members of this family were included in the genus TAXUS.

THE PODOCARP GENUS
(*Podocarpus* L' Her.)

Most of the needle-leaved members of the family are in this genus and, in some cases, as can be seen below, the needles are two to three inches long.

4.1 COMMON PODOCARP (*Podocarpus Nagi* Makino)

This is one of the most used (and overused) landscape specimens of the Gulf States, where homeowners often call it Podocarp*us*. Other species of the genus are also planted, but this one is ubiquitous. As with so many other "small" trees, the Podocarp has a tendency to outgrow the plans and wishes of its owner, but it takes well to judicious pruning.

These trees have enlarged and succulent seed-bearing stems which, like the CASHEW, is edible.

Two others, *Podocarpus dacrydioides* and *P. Totara* are valuable timber trees of New Zealand.

4.1

ARAUCARIACEAE

This small family from the southern hemisphere consists of about 30 species in 2 genera. It is seen planted-out in the southern U.S. and in greenhouses in the North.

THE ARAUCARIA GENUS (*Araucaria* Juss.)

Evergreen trees of this genus have branches that are regularly whorled. The scale-like leaves are spirally arranged on the stem.

5.1

5.1 NORFOLK-ISLAND PINE (*Araucaria excelsa* R. Br.)

A popular greenhouse plant in the North because of its symmetrical pyramidal form. It is also commonly planted in Miami, where it makes a striking tree if carefully fostered. Often the top is blown out by hurricanes and the tree itself starved and neglected until it becomes something less than beautiful. (Norfolk Island is near New Zealand but a part of Australia.)

BUNYA-BUNYA-TREE (*Araucaria Bidwillii* Hook.)

The foliage of this tree is much coarser than the species above and it suggests chain-mail armor. The seeds are edible. It is grown outdoors in the Miami area and in greenhouses in colder regions.

Other species of *Araucaria* (e.g., *A. Cunninghamii*) are important timber trees in Australia and New Zealand.

THE AGATHIS GENUS
(*Agathis* Salisb.)

The leaves of these evergreen trees are flat, broad, and leathery, and the seeds are winged on one side.

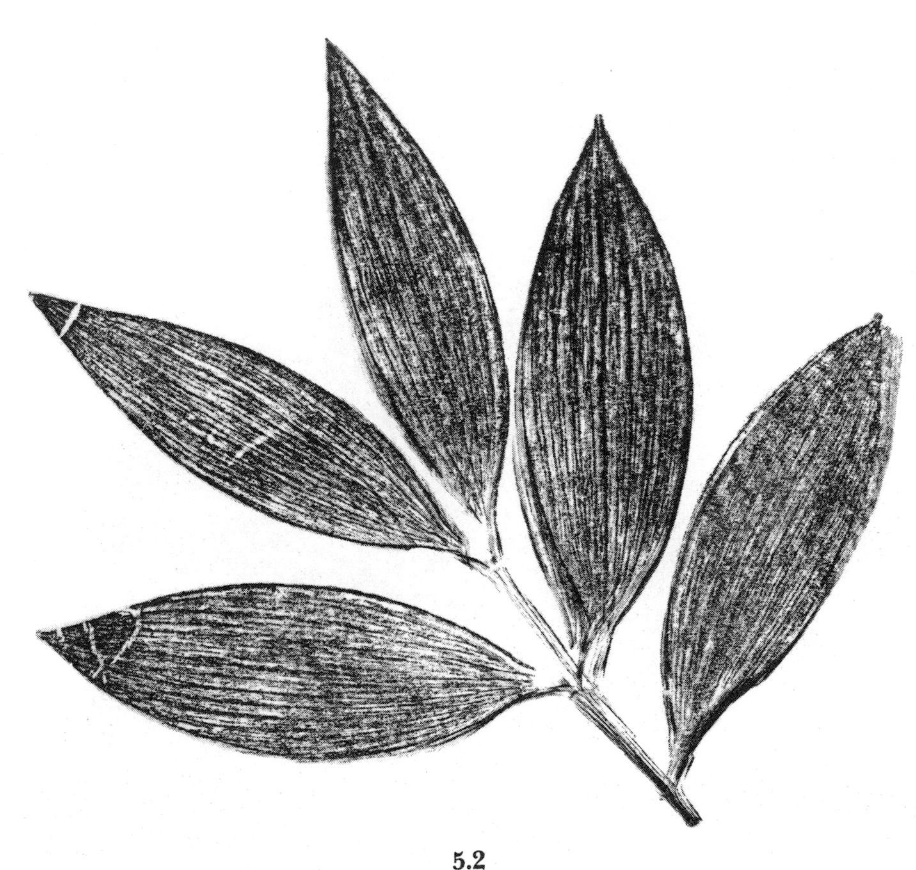

5.2

5.2 KAURI-PINE (*Agathis robusta* F. Muell.)

The print above was made from a swatch in a conservatory labeled as indicated. However, the leaves are larger than those of the *A. robusta,* which is used as a tub-shrub in Florida, so this is a moot identification. Bailey indicates that the leaves are narrow-elliptic.

The KAURI-PINE or DAMMAR-PINE is the source of one kind of dammar, a mixture of resins used in superior varnishes. The best commercial forms of these gums are those found as fossil deposits on ridges and in swamps and boggy ground. Lumps weighing up to 100 pounds are sometimes unearthed.

This and other species of *Agathis* (e.g., *A. Australis*) are important timber trees in New Zealand. In addition to being ornamental, their use in varnishes and for timber makes them a "three-product" tree.

According to Bailey this family consists of 9 genera and over 200 widely distributed species. By some authorities, however, the family is divided into three: (1) The true PINE family with resinous sap and scaly cones containing winged, wind-borne seeds; (2) The REDWOOD-CYPRESS family, including the BALD CYPRESSES with water-distributed cones; and (3) The ARBORVITAE-JUNIPER family culminating in the JUNIPERS, which have the cones modified into berries that are distributed by birds. We are here adopting Bailey's point-of-view.

THE PINE GENUS (*Pinus* L.)

This is the outstanding genus of the PINE family, both in the number of species and in economic importance.

Pine branches bear two kinds of leaves: small scale leaves and the foliage leaves, called needles. These occur in bundles (fascicles) containing 2 to 5 needles, although rarely only one needle or more than five needles may be present. Pine needles yield essential oils of use as disinfectants and in perfumery. All pines produce resins that are more or less medicinal. SLASH PINE (*P. caribea*) and GEORGIA or LONGLEAF PINE (*P. palustris*) are the chief sources of what are sometimes called "naval stores," the resinous sap or turpentine that is distilled to yield the liquid spirits of turpentine familiar as a paint solvent. Also recovered is the solid resin, called "rosin," which is an important ingredient of paints and varnishes, and is used for special purposes by violinists and wrestlers. Another product of this industry is the black pine tar or pitch.

Pine cones are much used in the decorative arts. Most of them open up when dry and ripe, usually on the tree, thus allowing the wind to transport the winged seeds. Some require the heat of a forest fire before they will open. In several species there is an umbo or sharp spine on each scale of the cone, possibly a device to keep squirrels from opening the cones and reaching the seeds before they are ready to take their chance in the gamble of seed dispersal.

The seeds of many of the species are minute, winged, and wind-borne, as noted above. Some, however, are large enough to be of interest to man as well as to the squirrel. Among these are *Pinus pinea,* the STONE PINE of the Mediterranean region whose seeds (pignolias) are cooked in soups and stews in Italy and elsewhere. They also have confectionary uses. In Mexico and southwestern U.S., the MEXICAN STONE PINE or PINON PINE (*P. cembroides*) is the source of an edible nut. *Pinus Lambertiana* and *P. monophylla* were important to the Indians and are still a small item of commerce although most of the so-called Indian nuts available today are from European sources.

Pine wood is called *deal* in the books, but I admit to never having heard this word with my own ears. The wood of the WHITE PINE (*P. Strobus*) of the northern forests was once one of the most important kinds and much of the beautiful furniture, paneling, and flooring that remains from the colonial period is made of it. However, after the eastern forests were sacked, the DOUGLAS-FIR (*Pseudotsuga taxifolia*) and others of the northwest took over the lead role.

Pines are important in home and park landscaping.

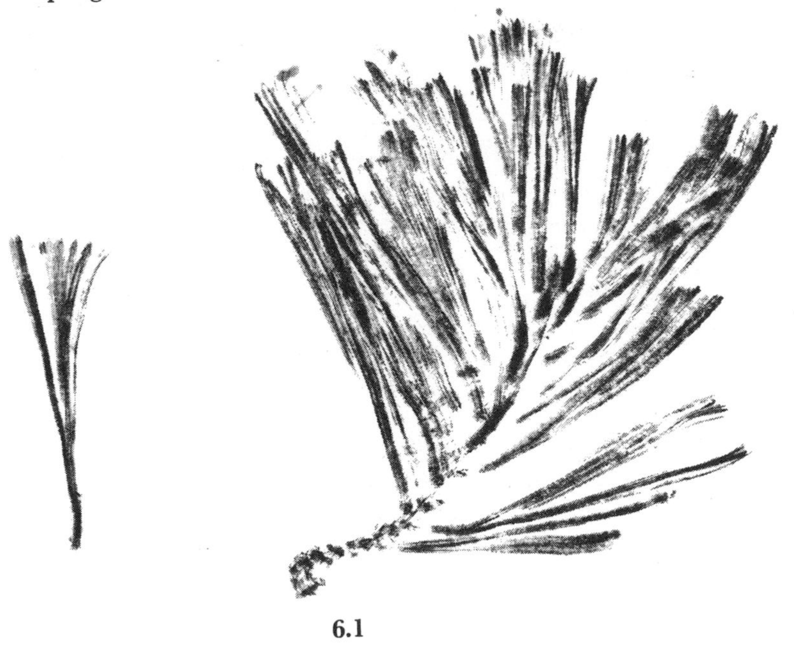

6.1

6.1 SCOTCH PINE (*Pinus sylvestris* L.)
Its needles are really in twos, but it insisted on moving during the printing process.

Also called SCOTS PINE, which is a Briticism, and SCOTCH-FIR, which is an error. This species is becoming increasingly popular for Christmas trees. The needles vary in length but are always short, for a Pine. Some are quite blue and all have more or less of a twist. The bark of this tree on trunks and branches about 8 to 10 years old is bright orange in color and has craft uses.

MUGHO or MOUNTAIN PINE (*P. Mugo* Turra.)

One of the mainstays of modern small-scale landscaping, some forms of this pine remain very dwarf while others grow large. There is a lack of consistency in nursery practice, much as with the JAPANESE YEWS.

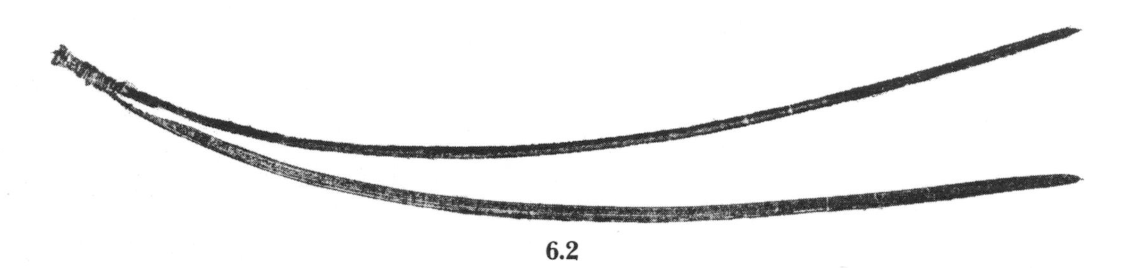

6.2

6.2 RED PINE (*Pinus resinosa* Ait.)

A native species of value both in forestry and horticulture. Sometimes called NORWAY PINE after a small town in Minnesota, but such names should be dropped. Its foliage is soft and usually pendulous, not stiffly bowed upward as in AUSTRIAN PINE. Other means of differentiation failing, see whether you can break a bunch of needles in two—easy with this species, difficult with the other.

AUSTRIAN PINE (*Pinus nigra* Arnold)

One of the largest of the Pines, making a handsome specimen tree in lawns and parks. Also used in forestry. The needles of this and other two-needle species can be made into chains by pulling out one needle and then tucking the end of the other into the empty socket. May be susceptible to pollution.

6.3 WHITE PINE (*Pinus Strobus* L.)

Has 5 flexible needles in the fascicle. Its resin is medicinal. The wood is easily worked and carved. Pitch boiled with equal parts of spruce gum is excellent for patching canoes.

6.4 LIMBER PINE (*P. flexilis* James)

The wingless seeds are dark brown and mottled with black.

THE CEDAR GENUS (*Cedrus* Loud.)

Four species of large evergreen trees have been assigned to this genus, but they are considered by some to be only forms of one species. The seeds have large membranous wings.

6.5 CEDAR-OF-LEBANON (*Cedrus libani* Loud.)

As the prints show, the evergreen True Cedars (*Cedrus*) and the deciduous Larches (*Larix*) have similar growth forms with the needles clustered on the spur shoots of the previous year and strung out along the shoots of the current year. Various other evergreens are called "Cedars," but we should show their lack of priority on the name with hyphens, as in Red-Cedar, White-Cedar, and so on.

The CEDAR-OF-LEBANON is renowned because of references to it in the Bible. It and two other species of *Cedrus* are attractive in cultivation.

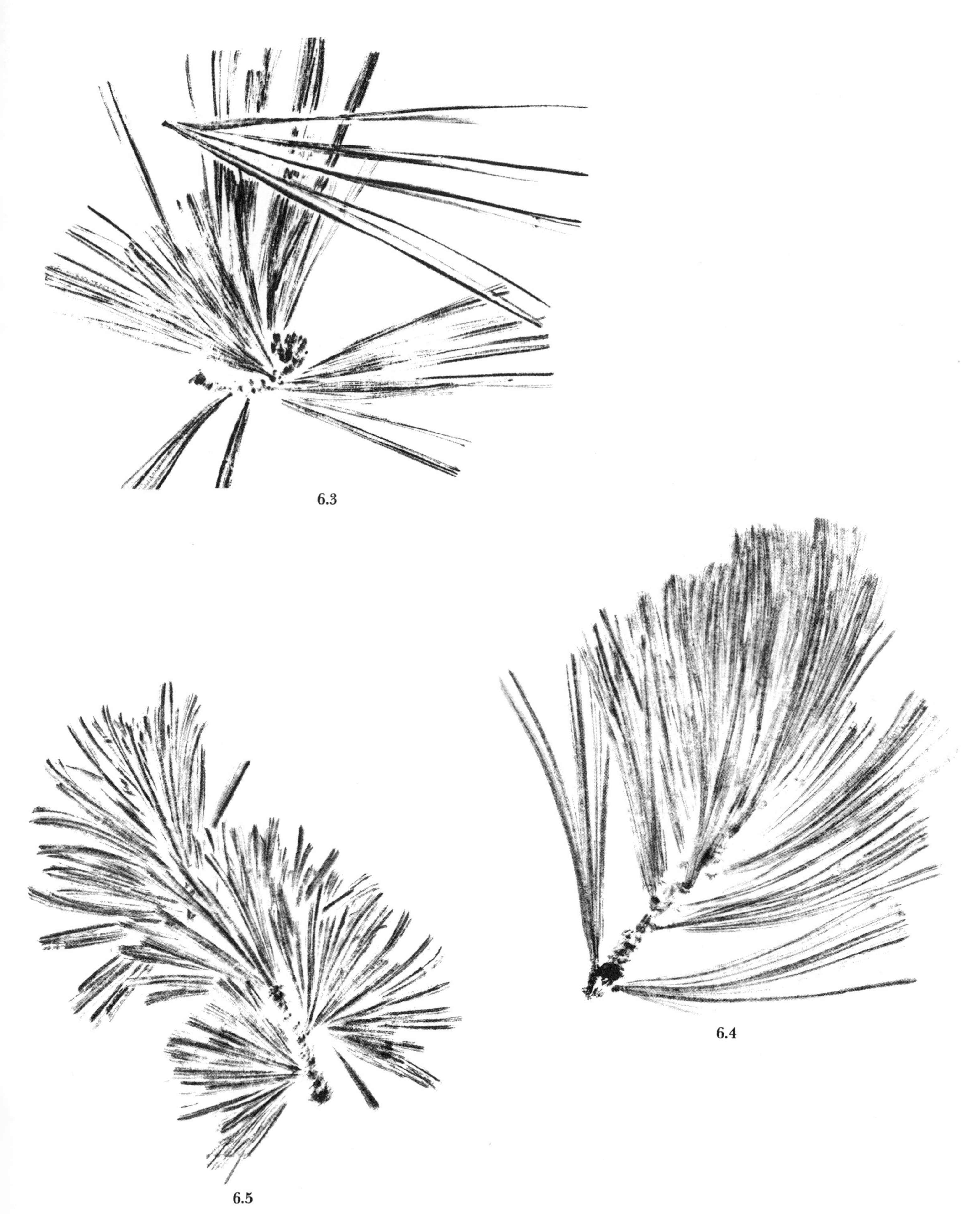

6.3

6.4

6.5

THE LARCH GENUS (*Larix* Mill.)

There are about 10 species of these deciduous, resinous trees native to the colder regions of the northern hemisphere. Their seeds have large thin wings and their leaves are borne in crowded clusters on short spurs.

The EASTERN LARCH or TAMARACK (*L. laricina*) and the WESTERN LARCH (*L. occidentalis*) provide some of the heaviest, strongest, and toughest of the softwoods. Because of its durability, it is also used in shipbuilding. The gum of the WESTERN LARCH has been used in the manufacture of baking powder.

6.6 EUROPEAN LARCH (*Larix decidua* Mill.)

The LARCHES and their kin all drop their leaves in the fall; I once knew of a Larch that was cut down in the winter by new tenants who thought it was a dead Spruce. This species is seen chiefly in parks, usually in the pendulous form. The AMERICAN LARCH or TAMARACK (*L. laricina* Koch) grows wild in northern swamps but may be successfully grown on drier land. Its roots make excellent cordage and the seeds, needles, and bark are eaten by several animals. The dark blue-green needles of *L. decidua* are not straight but have an unkempt look as compared with the regular look of *L. laricina*. Its cones are also much smaller. Both species produce timber and tannin.

Since the 1750s Venetian turpentine has been made from trees of *L. decidua* growing in the forests of central Europe.

6.6a

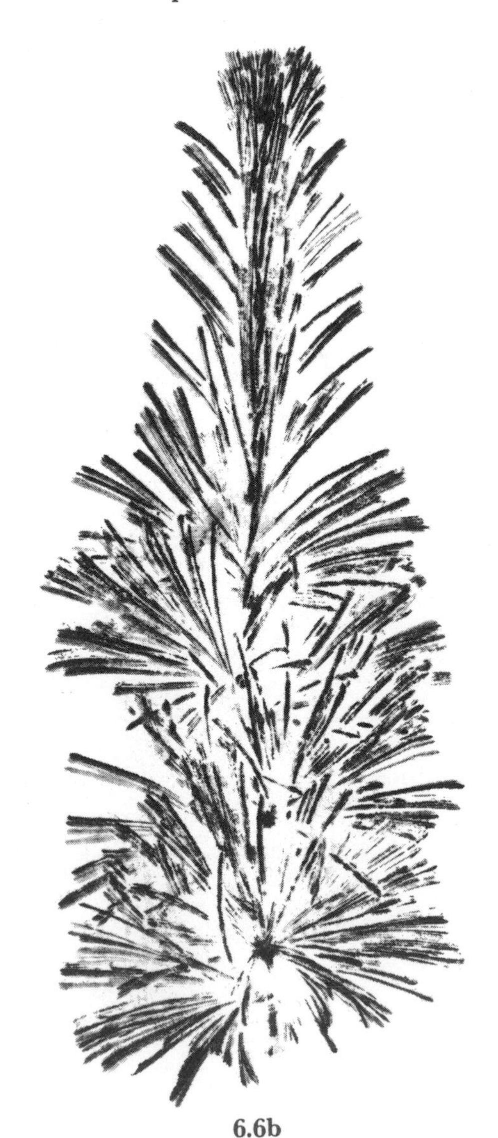

6.6b

THE FIR GENUS (*Abies* Mill.)

This genus of some 40 species has all the usual attributes of conifers: the wood is used for construction, paper, plastic, and so on, and the bark supplies tannin and drugs.

The FIRS, their close relatives the SPRUCES, and practically all northern conifers, have a pyramidal form, which gives them competitive power against other plants at the ground level and, by gradual diminution as the column grows, a maximum exposure to sunlight.

The evergreen conifers are exposed to transpiration all winter long, and it is quite probable that their roots extend below the frost level, otherwise they would dry out in the winter winds. There is a connection between this phenomenon and the advice of nurserymen in dry fall months to water evergreens into the winter. New transplants are shallow-rooted and hence do not reach below the frost level. Providing wind-screening may also be a necessity.

The cones of FIRS are held erect, instead of dangling as in the case of SPRUCES. The needles around the cones may be sharp-pointed as a protection for the seeds, but the ordinary needles do not have the sharp points that help the SPRUCES fend off browsers.

6.7

6.7 BALSAM FIR (*Abies balsamea* Mill.)

The wood of this important tree can be used in a variety of ways. In campcraft it is excellent for fire-by-friction (Mason). The blisters on the bark yield a medicinal resin (a true turpentine, not a balsam) once widely used for cementing lenses in optical systems. The foliage is used for stuffing souvenir pillows and in the days before air mattresses, "browse" shoots were cut for camp bedding (Kephart). The roots can be split to provide cordage and as a Christmas tree it holds its needles.

CONCOLOR FIR (*Abies concolor*) is used for landscaping and is often mistaken for BLUE SPRUCE, but the needles are longer, flatter, and not sharp-pointed.

NORDMANN FIR (*Abies Nordmanniana*) is as dark as the CONCOLOR FIR is light, the two together giving a notable contrast.

SOUTHERN FIR (*Abies Fraseri*) is of interest geographically on account of its limited range in the Smoky Mountains and because there is a prostrate form of it.

THE SPRUCE GENUS (*Picea* A. Dietr.)

There are approximately 40 species in this genus, and because the two genera are closely related most of the notes under the Firs apply to the Spruces also.

The SPRUCE forests of Canada have long supplied most of the pulp for newspapers, although they are also in competition in this regard with the PINES of our southern states.

SPRUCE cones assume a dangling position on the tree. They are useful for winter decoration and novel in that the scales close when wet and open when dry. The seeds are wind-borne.

The needles are sharp-pointed, sometimes painfully so. This device is aimed not against the tender skin of man but the tongues and gums of browsing animals, especially deer. In this sense there is a parallel between these sharp-pointed needles and the spines of cacti, since both offer protection to plants that may be the only food available in their respective habitats.

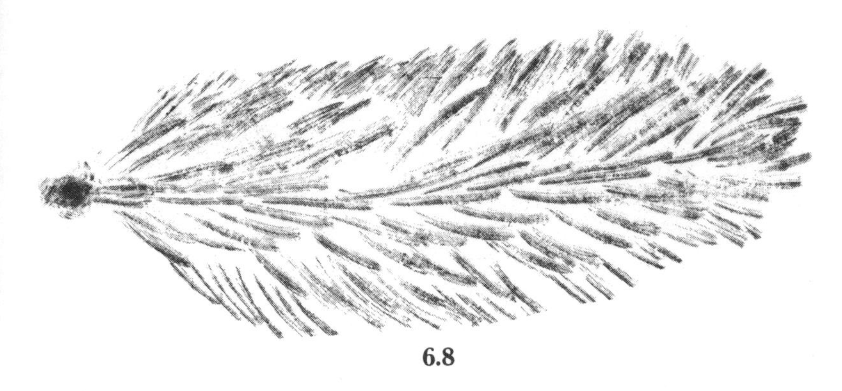

6.8

6.8 NORWAY SPRUCE (*Picea Abies* Karst.)

This Old-World species and the native WHITE SPRUCE (*P. glauca*) and COLORADO (BLUE) SPRUCE (*P. pungens*) have been much planted for ornament, especially for green or whitish winter effects. They have also come into considerable vogue for shelter belts. This is reminiscent of the pioneer days when they were planted around farmhouses and plots as snow fences and wind screens.

RED SPRUCE (*P. rubens*) and the BLACK SPRUCE (*P. mariana*), native spruces of the American northeast, entered into the folklore of that region by producing a sort of chewing gum purportedly enjoyed by young and old alike. My own experiences with it, based on a packet purchased in Maine, were not too happy. The roots of these trees when split make excellent lashing cord.

THE HEMLOCK GENUS (*Tsuga* Carr.)

A genus of 10 species of evergreen trees native to the temperate zones of North America, Japan, and China.

6.9 CANADA HEMLOCK-TREE (*Tsuga canadensis,* Carr.)

A native evergreen conifer and one of the most graceful trees when growing healthily in a favorable environment. It takes well to shearing (if you can stand such brutality) and can be used for formal specimens and hedges. The tiny cones attract winter birds. The bark is more important commercially than the wood, being an important source of tannin for shoe leather and drugs.

T. canadensis is the source of one of the better beverages that the woodsman can obtain from field material. Steep a half-dozen branchlets with foliage in a small pot of boiling water for 5 minutes. Remove the twigs and sugar the resulting "Hemlock tea" to taste.

Not good as a Christmas tree since the needles fall when dry.

6.9

THE PSEUDOTSUGA GENUS
(*Pseudotsuga* Carr.)

Only 5 species of evergreen trees are included in this genus. The leaves are grooved above and have white bands on either side of the mid-rib beneath. Oval scars on the otherwise rather smooth branchlets indicate where leaves had previously been attached.

6.10

6.10 DOUGLAS-FIR (*Pseudotsuga taxifolia* Britt.)

This native of the Pacific Northwest is now perhaps the most important timber tree in the world. It is used in afforestation in England and other countries. By applying the most advanced forestry techniques, the lumber industry is attempting to assure a continuous supply. The Rocky-Mountain form of this tree, with blue foliage, is often planted eastward. The bracts of the cone protrude from beneath the scales like little fish-tails.

Some authorities include this family with the Pines. The cones are very woody with thick, wide-spreading scales that have no distinct bracts. When considered a separate family, there are some 8 genera and 15 species assigned to it.

THE TAXODIUM GENUS
(*Taxodium* Rich.)

A small genus of three species native in eastern North America and Mexico. They are needle-bearing, and deciduous by virtue of the shedding of small branchlets.

7.1a and 7.1b BALD CYPRESS or "CYPRESS" (*Taxodium distichum,* Rich.)

Perhaps the most famous tree of the Deep South. It is especially known in the North for the "knees" used for lamp bases, wall brackets, and in various other decorations.

The wood is highly resistant to weather damage and is used for bird houses, shingles, and other outdoor construction. The heart-wood has a rancid odor but no taste. The fibrous bark appears suitable for insulation. Doaty and vermiculate sections of the wood have been used for paneling. The needles turn red and will provide a red dyestuff.

The male flowers are very conspicuous and pollination is by the wind. The cones are disseminated chiefly through floating. The tree form is columnar.

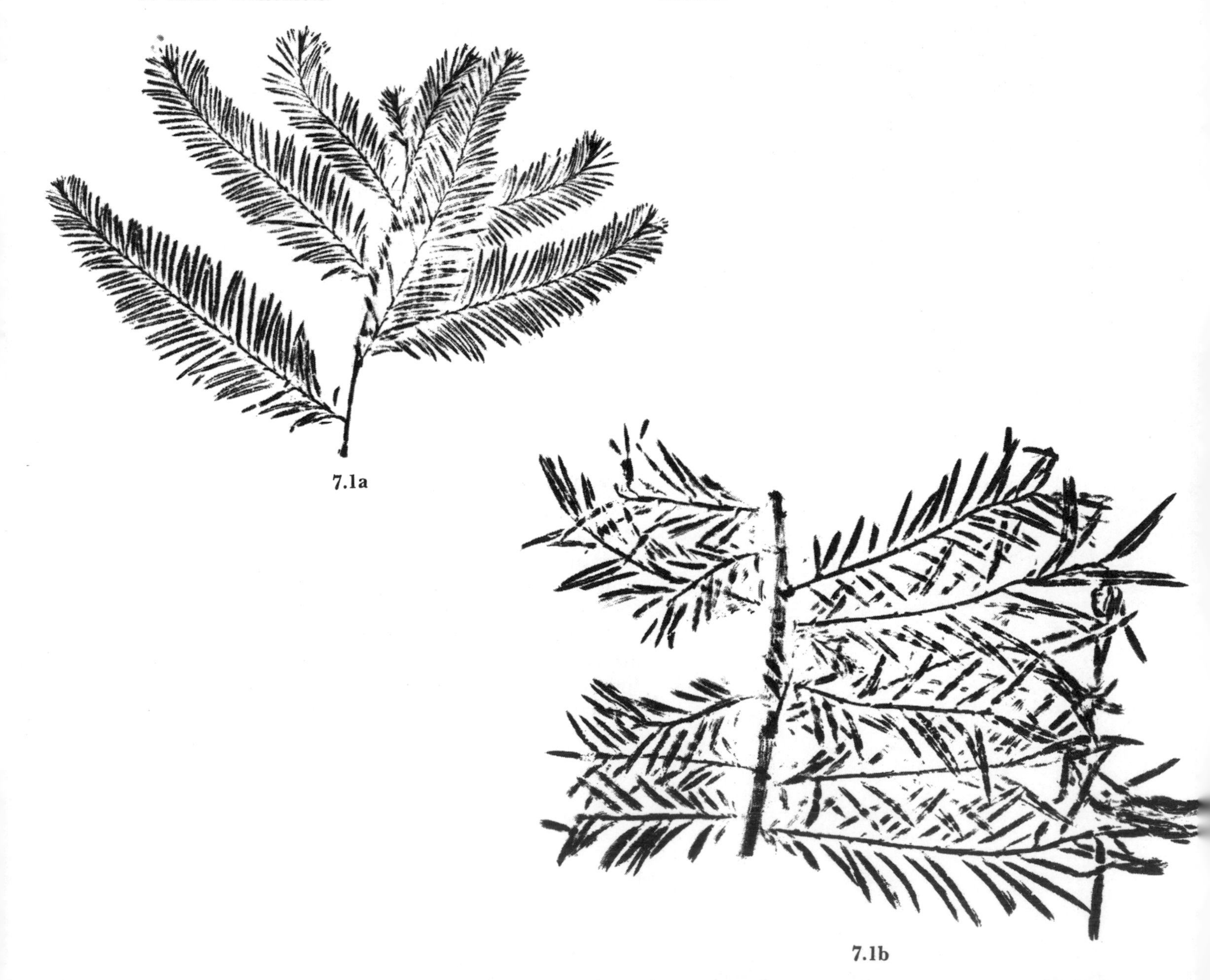

7.1a

7.1b

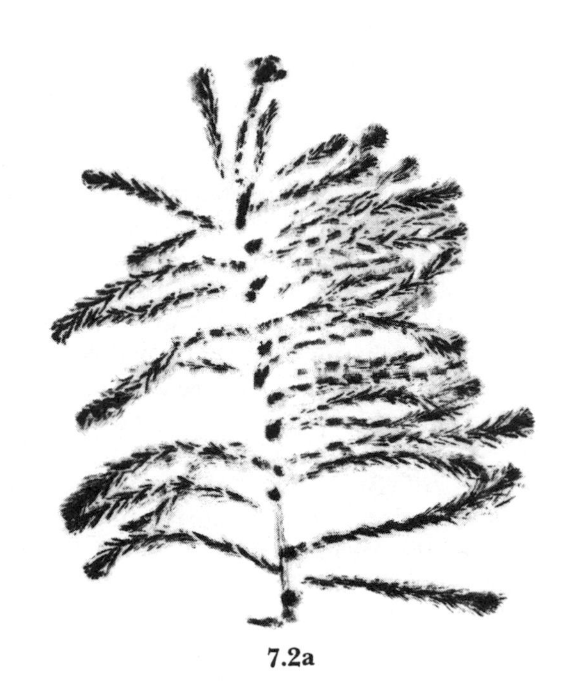

7.2a

7.2a and 7.2b POND CYPRESS (*Taxodium ascendens* Brongn.)

A smaller tree than its relative on the previous page and not nearly so well known. In Spring Grove Cemetery, Cincinnati, a pond has been surrounded with it, but other Cincinnati locations have been destroyed. In addition to the marked foliage difference, this tree differs greatly in form. It is truly pyramidal and not columnar as in the BALD-CYPRESS. I have not seen "knees" produced by it. Fernald and others do not consider it distinct.

7.2b

ABOUT CYPRESS KNEES

For centuries people have been saying that these odd structures developed as aeration devices for conveying oxygen to the submerged roots. Such devices do exist in the WATERLILIES and, perhaps, the aquatic CROWFOOTS, but very doubtfully, to my mind, in the BALD-CYPRESS. Here we have very solid (not porous) wood covered by a quite impervious bark.

By way of explaining what I believe to be the real "purpose" of BALD-CYPRESS "knees," let me call your attention to the anchoring function of special roots of the various trees known as MANGROVES, of CORN, and of the SCREW-PINES (*Pandanus*). All of these, and many other plants growing in water or on land tend to produce prop roots, which anchor them against the wind.

Why could not CYPRESS "knees" also be an anchoring device? If you can find a well-developed but isolated tree growing in the water, observe how the knee system would help prevent the tree, shallow-rooted as are all aquatic trees, from being toppled over by strong winds such as hurricanes.

7.3 DAWN-REDWOOD (*Metasequoia glyptostroboides*)

One of the several sib-pines discovered in China as late as the twentieth century.

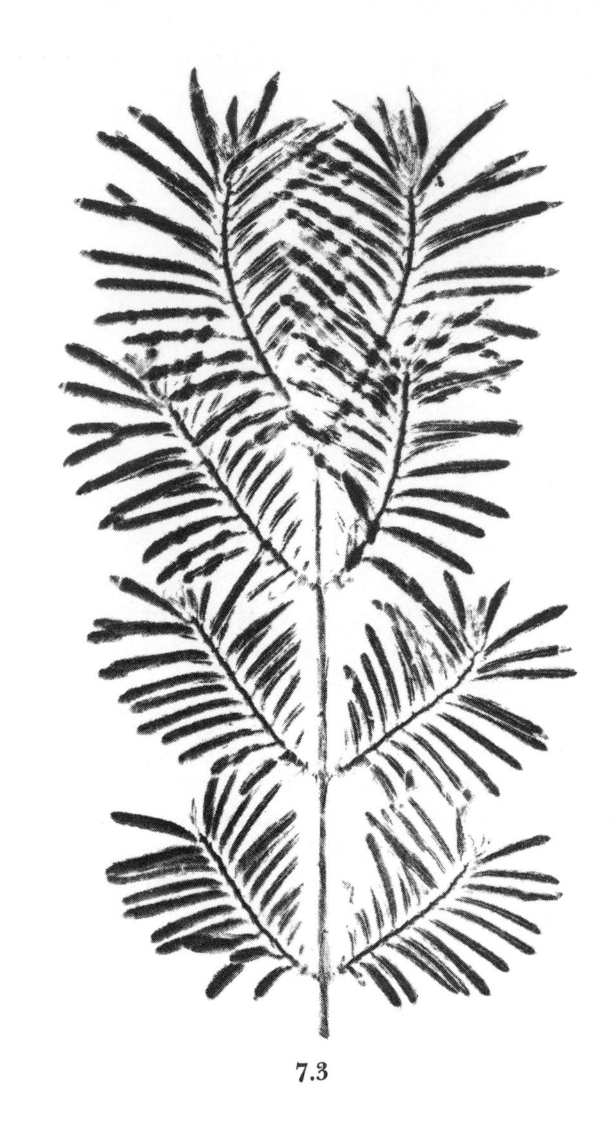

7.3

GLYPTOSTROBE-TREE (*Glyptostrobus pensilis*)

Another relatively recent discovery in the Orient of members of the REDWOOD group. Apparently it is not so hardy as the DAWN-REDWOOD and in this country is planted only in the Deep South.

(COAST) REDWOOD (*Sequoia sempervirens* Endl.)

This is a bulkier or heavier tree than the BIG-TREE (below), if not so tall. It has long been a bone of contention between conservationists and would-be ravagers. The wood is so valuable that it is imitated by staining others. The bark has been used for insulation. Burls from this tree will sprout beautifully if placed in water and are occasionally an item of sale.

BIG-TREE (*Sequoiadendron giganteum* Buchholz)

Probably the tallest growing tree in the world, surpassing 350 feet, and rivaled only by some of the Australian EUCALYPTUS. Its name is associated with the names of such famous naturalists as John Muir and Enos Mills.

A family of about 15 genera of resinous trees and shrubs valuable as ornamentals and for timber. The fruits vary from woody or leathery to berry-like cones. There are some 125 species.

THE CYPRESS GENUS (*Cupressus* L.)

Trees of this genus exhibit long shred-like plates of bark that are rather persistent. Some species, however, possess thin barks of curling nonfibrous plates that separate from the trunk annually. There are more than 20 species in the genus.

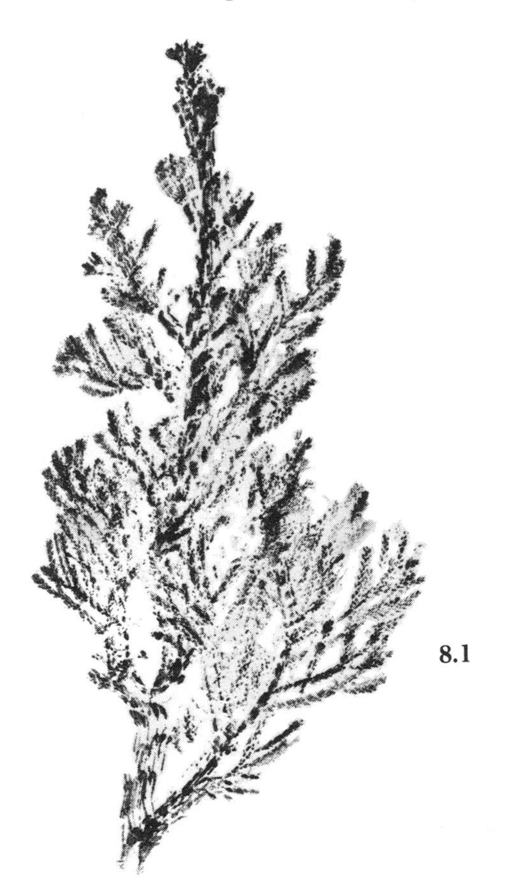

8.1

8.1 ITALIAN CYPRESS (*Cupressus sempervirens* L.)

This is the CYPRESS of literature associated with cemeteries. Its remarkably narrow form is useful in landscaping, especially for "tight places."

THE ARBOR-VITAE GENUS (*Thuja* L.)

Six species of aromatic resinous trees whose branches are flattened and frond-like make up this genus. The wood is soft, light, and easily worked, and has been used for boat and canoe building and for shingles.

8.2 AMERICAN ARBOR-VITAE (*Thuja occidentalis* L.)

A large tree of pyramidal habit but cultivated in a variety of forms including a flat sphere called GLOBE ARBOR-VITAE and a columnar form miscalled PYRAMID ARBOR-VITAE. *Thuja occidentalis* is also miscalled WHITE CEDAR and the so-called cedar posts of the upper Middlewest are made from its wood. Large wild trees have an intricate trunk structure. The bark is used by craftsmen; the foliage is used by florists and is medicinal. Young specimens are favorite signal posts of dogs.

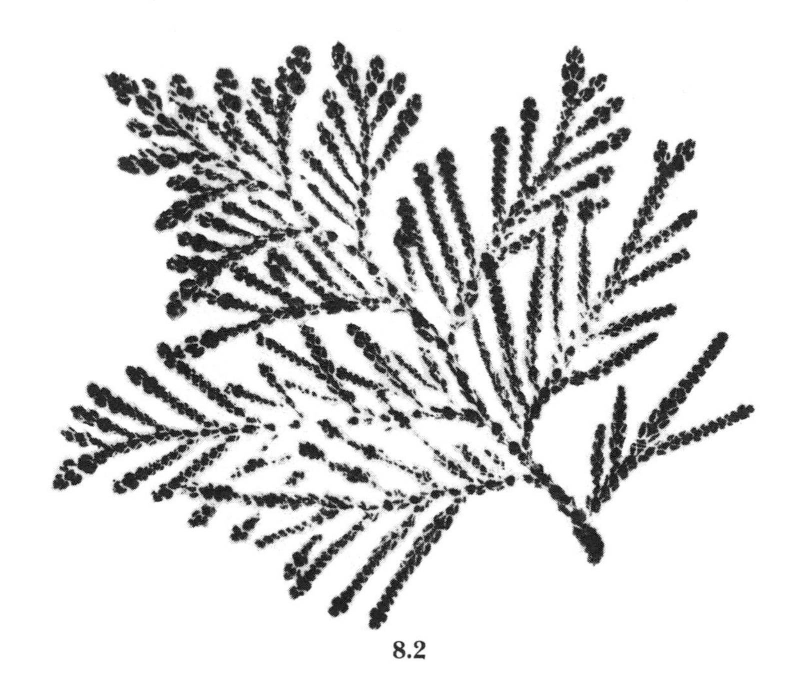

8.2

THE FALSE CYPRESS GENUS
(*Chamaecyparis* Spach.)

A rather small genus consisting of six species of evergreen trees with flattened branchlets and scale-like leaves.

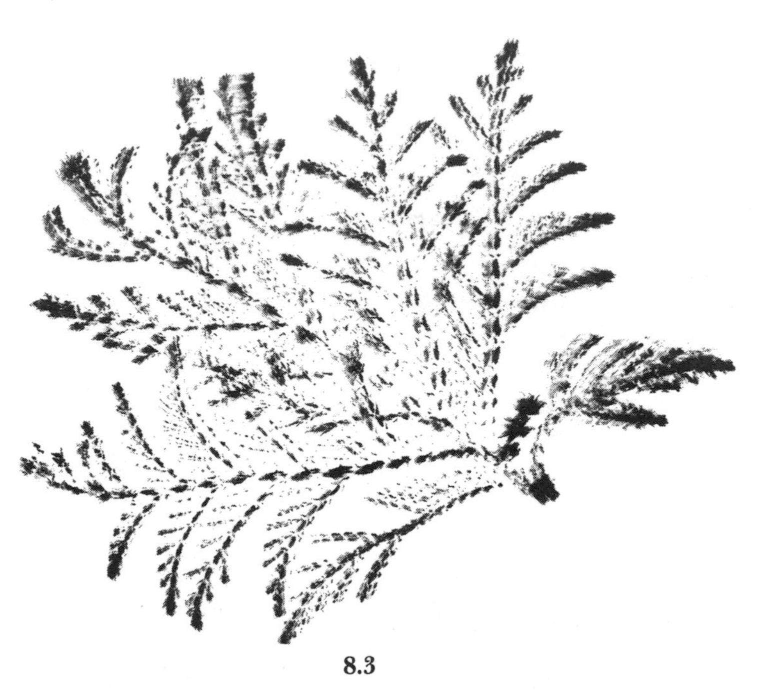

8.3

8.3 SAWARA CYPRESS (*Chamaecyparis pisifera* Sieb. & Zucc.)

Has many cultivated varieties including those with golden-yellow foliage (var. *aurea*), gracefully pendulous branches (var. *filifera*), etc.

THE INCENSE-CEDAR GENUS
(*Libocedrus* Endl.)

The species shown below represents one of the eight species of evergreen resinous trees included in this genus.

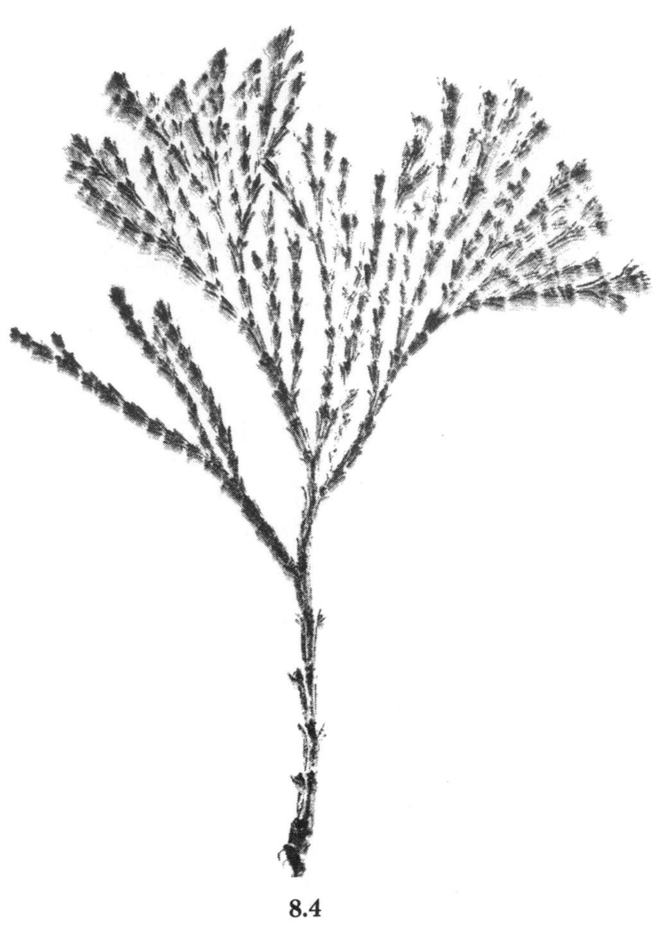

8.4

8.4 INCENSE-CEDAR (*Libocedrus decurrens* Torr.)

A tree with bright cinnamon-red bark and scale-like leaves whose bases extend down the twigs. The leaves are lustrous green on both sides. Sometimes growing to 200 feet, they are important lumber trees of western U.S. and are used eastward in landscaping. The wood has been popular for the manufacture of pencils and Venetian blinds. The foliage is deliciously fragrant when crushed.

THE JUNIPER GENUS (*Juniperus* L.)

A group, some 60 species, of very hardy (and a few subtropical) evergreens mostly of small stature. In addition to those of tree form, there are several mat-type species and varieties adapted to mountain and shore habitats.

The leaves, as best exemplified by the RED-CEDAR, are of two types, called juvenile and adult. The juvenile leaves are awl-shaped and are found on young specimens where they are a protection against browsers. The adult leaves are mere scales appressed closely to their stems and are found above the reach of deer and other browsing animals. Their dense structure and hard surface are helpful in conserving water during the "physiological drought" of winter.

The bark of both the Red- and White-Cedar has provided pioneers with fiber for making cord, and the split roots, called by the Indians "Watap," served as cord and lashing material.

8.6

8.5 COMMON JUNIPER (*Juniperus communis* L.)

Not often found growing wild in the middle west, where the RED-CEDAR abounds. The berries are edible in an emergency. They are of some importance as an antiseptic diuretic, and as a spice used in the flavoring of gin and other alcoholic beverages. They are occasionally seen on the spice shelves of stores. The sprigs were once used for the smoking of German Westphalia hams and the foliage is burned as incense in India.

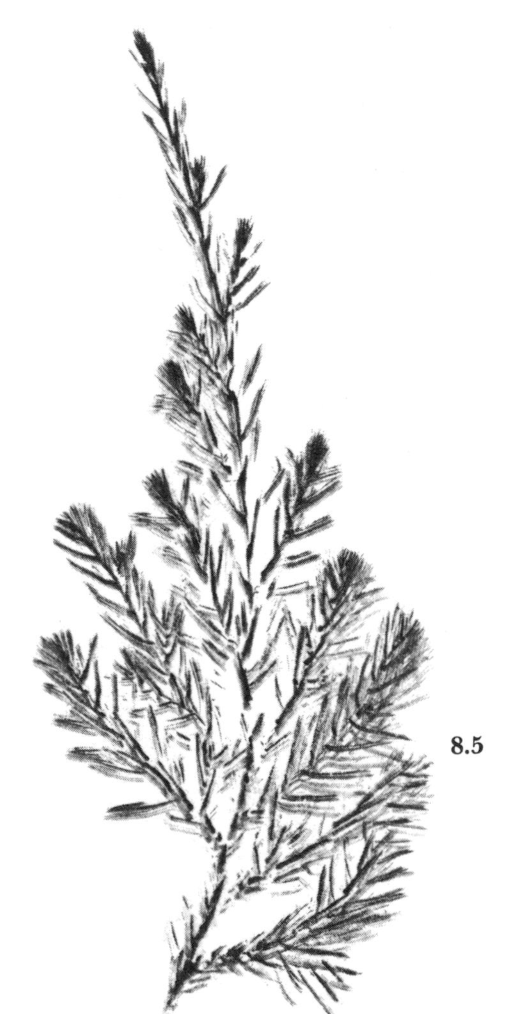

8.5

Cultivated in a variety of forms including prostrated (var. *depressa*) and narrow-upright (IRISH JUNIPER) and with nodding tips (SWEDISH JUNIPER). The vase-form is perhaps the most attractive.

8.6 RED-CEDAR JUNIPER (*Juniperus virginiana* L.)

Commonest evergreen of the limestone middle west. The pinkish, aromatic wood has an odor agreeable to man, but somehow an insectifuge to clothes moths. This wood is made into "cedar chests," and the extracted oil is used in floorsweeping compounds. Some 68 species of birds are known to eat its berries, although the seeds are not digestible. Seedling trees are therefore frequently seen along fence lines. Silver, blue (or white) forms are common in cultivation. The CANNAERT (CANNART) JUNIPER is a dark green, open-growing form that produces an amazing abundance of blue berries.

"CEDAR-APPLES" are an alternate-host fungus that spends part of its life-cycle on WILD CRAB-APPLE trees. The spores are distributed by flies.

There are over 300 species distributed within three genera in the Willow-Poplar family. They are rapid-growing, soft-wooded plants whose flower is a catkin (as in the Pussy Willow). In most members of the family the flowers occur in early spring before the leaves appear. The flower-bud has only one scale.

Of the three genera, only the Willows and the Poplars are important. The POPLARS may be upland trees although they also frequent the bottomlands. They have broad leaves. The waterside WILLOWS usually have narrow leaves in common with many tropical plants subject to flooding (Merrill).

Both POPLARS and WILLOWS produce tufted seeds, which are disseminated by the wind; the POPLARS are also pollinated by the wind, being tall trees suitable to it. But the lower, wetland WILLOWS, in the life-zone of insects, are insect-pollinated. Again, the PUSSY WILLOW is a good example.

The POPLARS are subdivided into COTTONWOODS, with resinous buds that supply medicines to man and propolis to bees, and the ASPENS, whose buds are not sticky. Some of the ASPENS have bark almost as white as some BIRCHES, and this may be also a matter of camouflage to protect the tree during the winter against bark-eating deer and other cervines.

The wood of some WILLOWS is made into charcoal for gunpowder and water filters. Mason lists SILVER WILLOW as among the best for fire-by-friction. WILLOW bark was the original source of medicinal salicylic acid, which is a cognate of aspirin (acetylsalicylic acid), now synthesized. Note the "sal" of *Salix* here.

The aesthetic or landscape value and the conservation uses of WILLOWS are also important.

THE WILLOW GENUS (*Salix* L.)

Most of the over 300 genera included in this family are members of the WILLOW genus. In general they are trees and shrubs but in some habitats prostrate forms are prevalent.

WILLOWS are commercially useful in two respects. The bark, especially that of BLACK WILLOW (*S. nigra*) provides tannin for light-colored, soft, glove leather. It is much favored in the U.S.S.R. Second, the resilience of the wood lends itself to basket-making, wicker-work, and the manufacture of furniture, baseball bats, and excelsior.

A tea made from the leaves and bark of various WILLOWS has been used in Appalachia to break up fever.

9.1 BLACK WILLOW (*Salix nigra* Marsh.)

A native species, often broadly V-shaped. Note the stipules.

9.2a and 9.2b EUROPEAN PUSSY WILLOW (*Salix caprea* L.)

Called GOAT WILLOW in the books. This and not the native *Salix discolor* is *the* PUSSY WILLOW of yards.

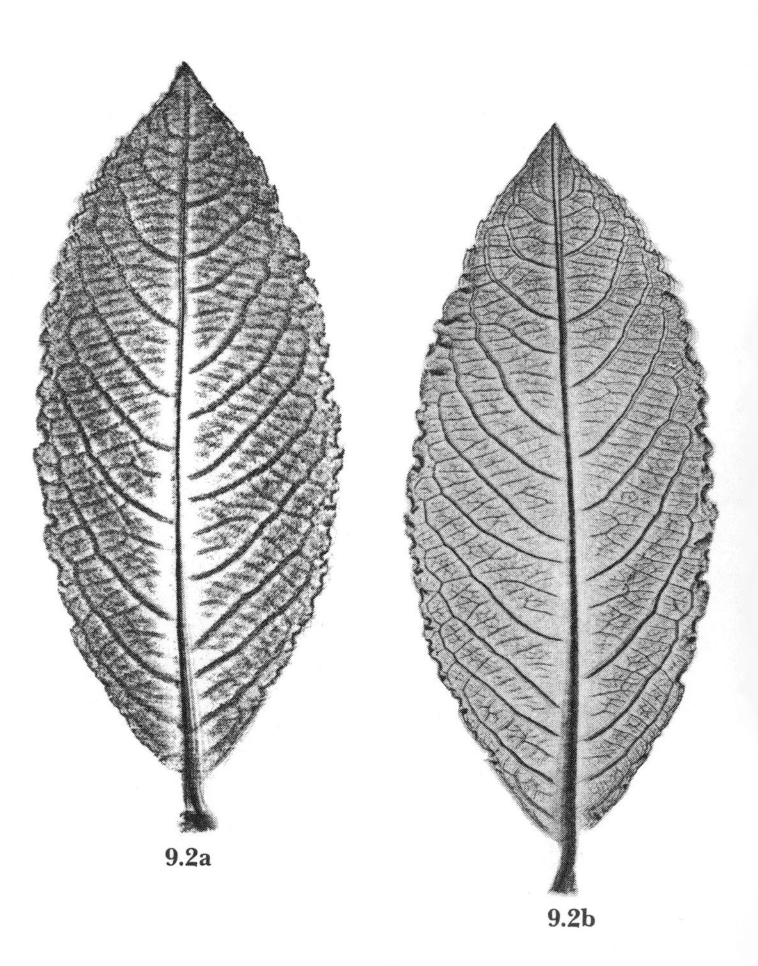

9.2a 9.2b

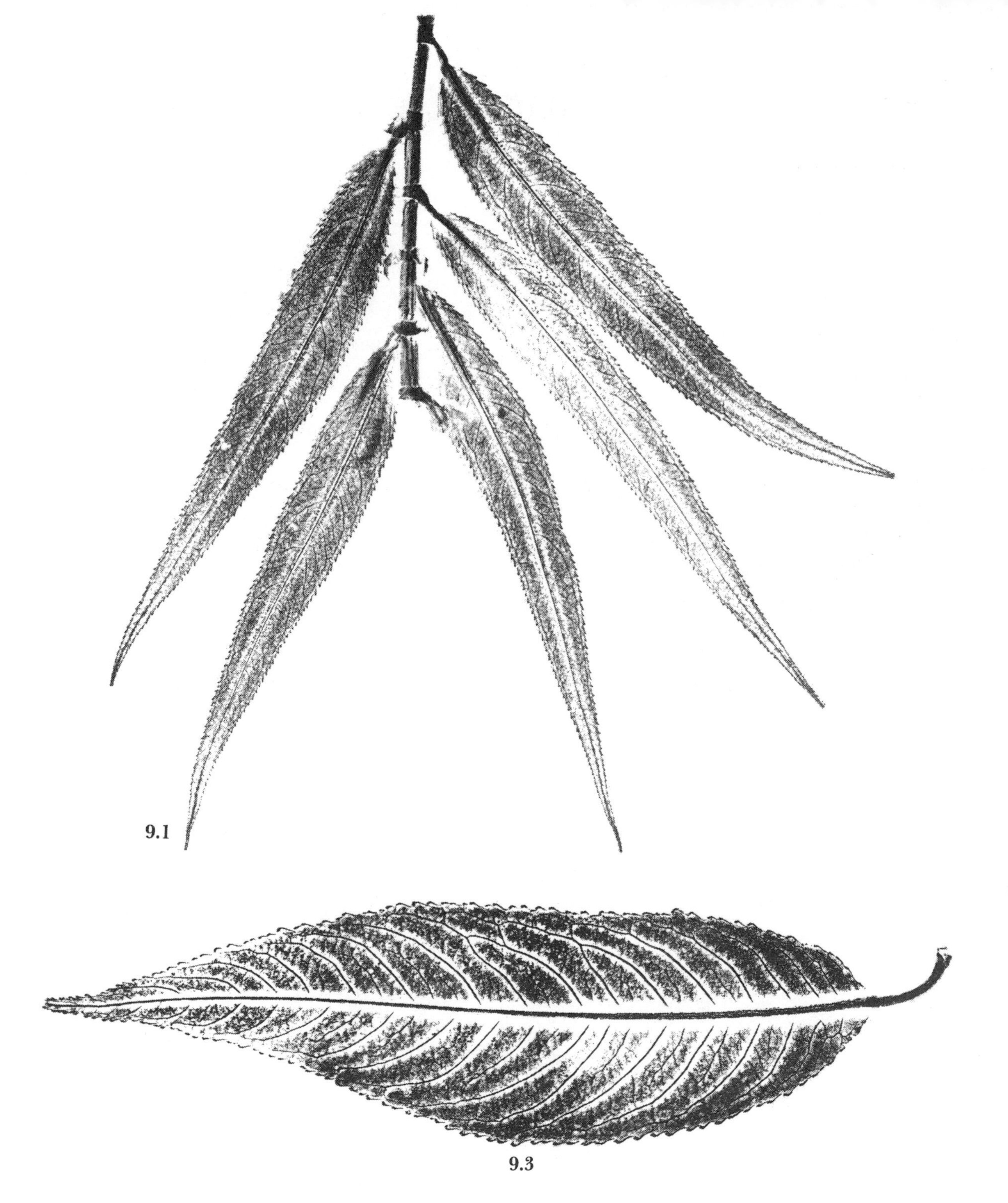

9.1

9.3

9.3 CRACK WILLOW (*S. fragilis* L.)

A rapidly growing tree to 90 feet, planted for shade and as a hedge. Branchlets break and fall freely.

9.4 RED WILLOW (*S. rubens* Schrank)

Has the brittle branchlets of CRACK WILLOW and the leaves white beneath as in WHITE WILLOW.

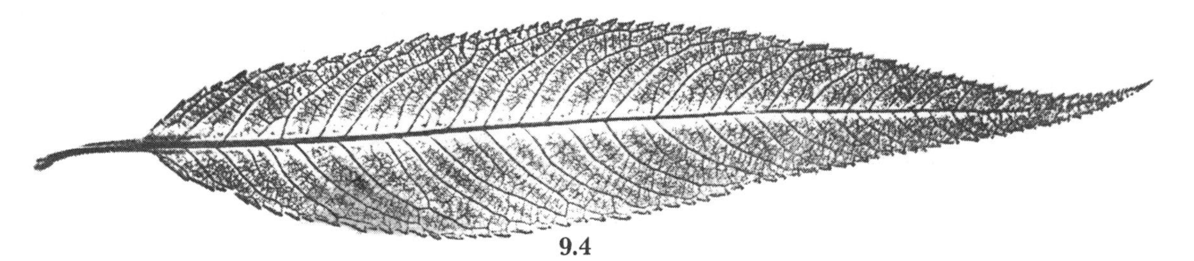

9.4

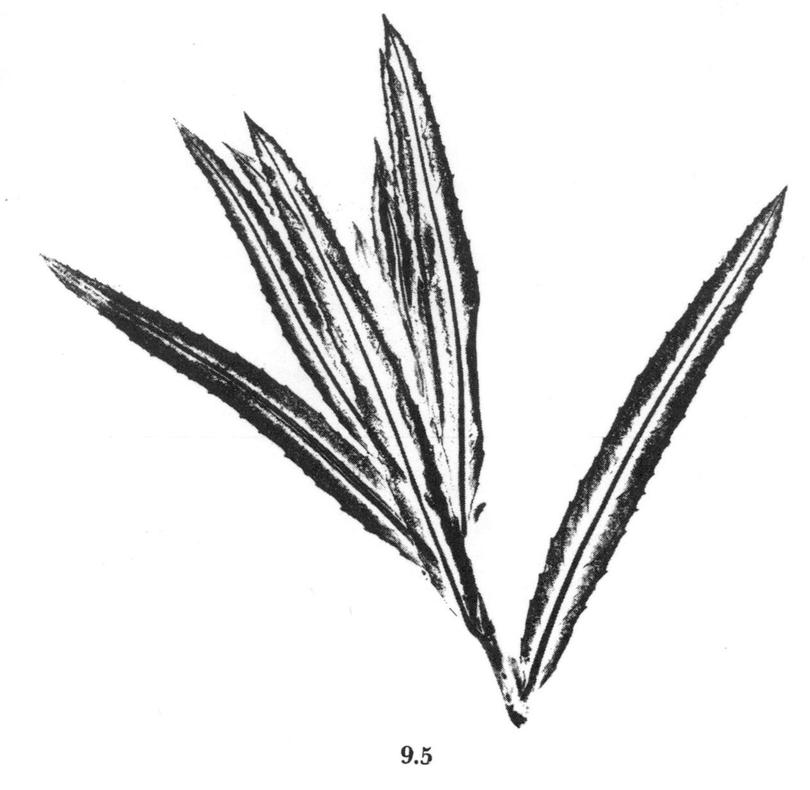

9.5

9.6

9.7

9.5 SANDBAR WILLOW (*Salix interior* Rowlee)
Often pioneers on sandbars and beaches.

9.6 SILKY WILLOW (*S. sericea* Marsh.)
Another brittle-based willow. A shrub growing to about 12 feet, usually on banks of streams.

9.7 ARCTIC WILLOW (*S. Elaeagnos* Scop.)
A native of Europe and Asia Minor, but very common in cultivation here.

9.8 PURPLE OSIER (*Salix purpurea* L.)
A small shrub whose bud scales are purple-colored. Planted as an ornamental and sometimes used in basketry and for making an Indian "willow-bed." See Jaeger.

9.9 YELLOW WILLOW (*S. alba* var. *vitellina* L. Stokes)
This tree, which grows to 75 feet, has been naturalized from Europe and is now one of the commonest varieties in the U.S. The branchlets are yellow and without hairiness.

9.10 WEEPING WILLOW (*S. babylonica* L.)
A frequently planted tree that does best in lowlands.

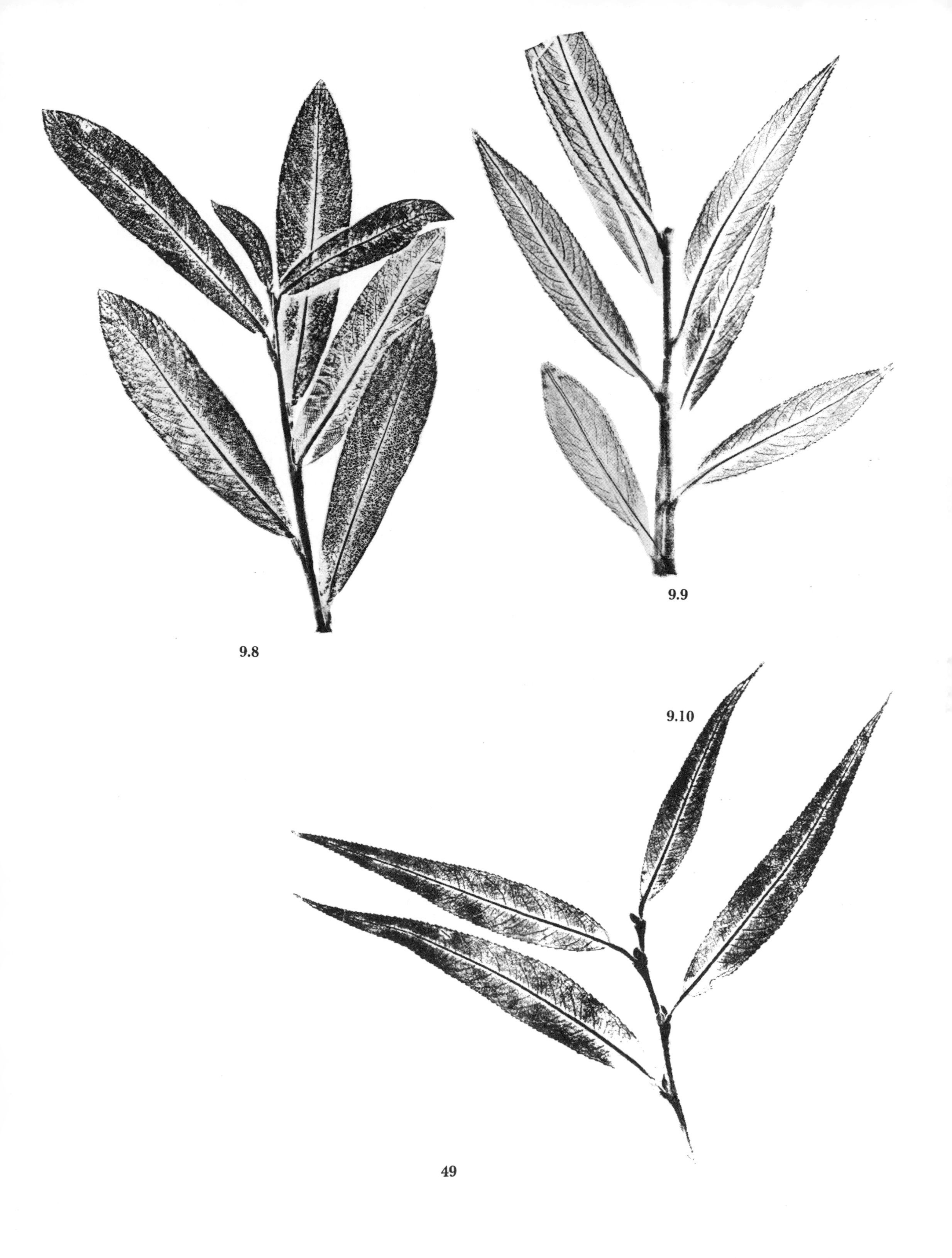
9.8
9.9
9.10

THE POPLAR (ASPEN, COTTONWOOD) GENUS (*Populus* L.)

The three common names above apply to some 30 to 40 species in this genus. In general, the POPLARS have sticky, aromatic buds, while the ASPENS do not. Conversely, the latter have vertically flattened leaf-stalks while the former do not. The COTTONWOODS derive their name from the seeds that float on the air supported by wisps of "cotton."

Since POPLAR wood is soft, light, and even-grained, it is very easily worked and has been widely used in the manufacture of boxes, excelsior, and paper for books. It rots very quickly when wet.

POPLARS have several bud scales as opposed to the WILLOWS, which have only one.

9.11 BIGTOOTH ASPEN (*Populus grandidentata* Michx.)

Found more frequently to the south of the range of *P. deltoides*. Leopold reminds us that the ASPENS provide food for grouse in winter.

9.12 SILVER or WHITE POPLAR (*P. alba* L.)

The leaves are white beneath and glossy green above.

9.11

9.12

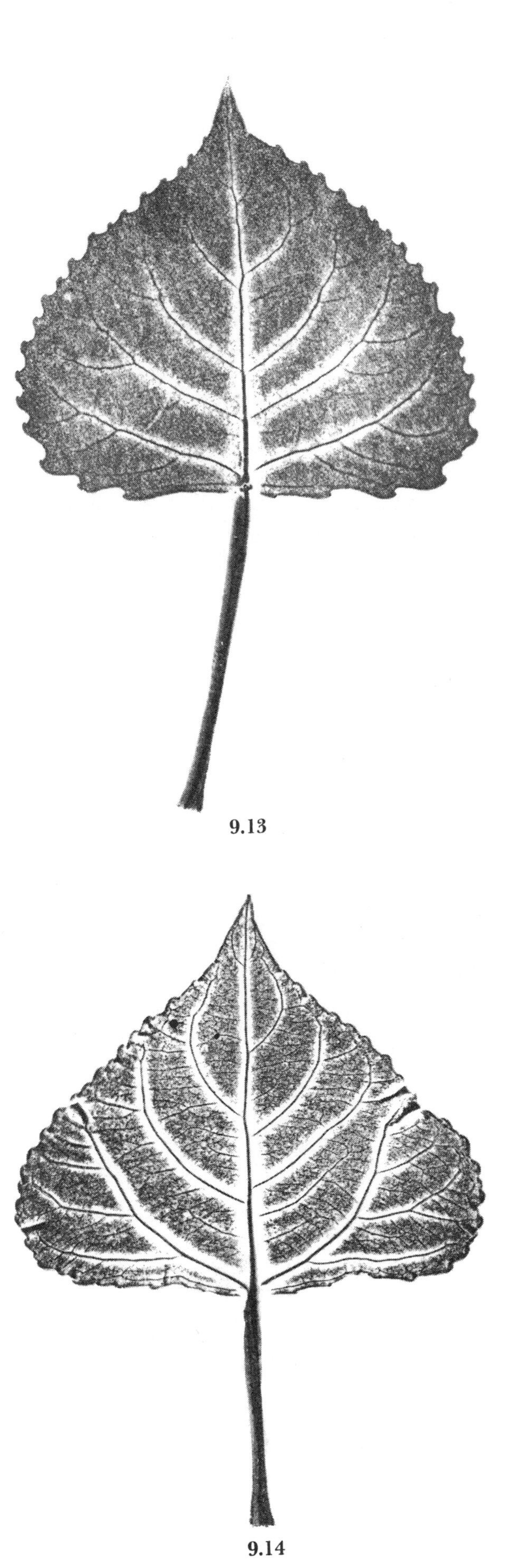

9.13

9.14

9.15

9.13 COTTONWOOD (*Populus deltoides* Marsh.)

The common wild POPLAR of the eastern half of the U.S. The bark is unusually thick. Larger leaves have a truncate-cordate base.

9.14 LOMBARDY POPLAR (*P. nigra* var. *italica* Muenchh.)

Familiar for its remarkably narrow form. It is a short-lived tree.

9.15 BALM-OF-GILEAD (*P. candicans* Ait.)

The leaves are pale beneath. Has been used in the production of cough medicine.

This family consists of two genera and about 50 species of evergreen and deciduous shrubs. The leaves are usually aromatic, the fruit a one-seeded drupe or nut often with a waxy coat.

10.1 BAYBERRY or CANDLEBERRY (*Myrica pensylvanica* Loisel.)

The BAYBERRY is an ornamental shrub that thrives in poor soil. It is frequent in both dry and wet soils along the eastern coast, south to North Carolina and inland to Ohio. Where it grows abundantly, the berries are gathered, boiled to separate the wax from the seed, and the wax is made into fragrant candles. The berries are sought after by myrtle warblers, crows, and other birds. The fruit and leaves are used for flavoring, the dried leaves as a tobacco substitute.

WAX-MYRTLE (*M. caroliniensis*)
Also produces waxy berries.

SWEET GALE (*M. Gale*)
The leaves have been used for flavoring meat. Its berries have no wax.

10.2 SWEET-FERN (*Comptonia peregrina* Coult.)

This aromatic shrub is in no way related to the true ferns but merely bears a superficial resemblance to them. At one time there was but one relict stand of SWEET-FERN in all north-central Illinois. Perhaps it should be planted more and the soil modified to suit it.

A decoction of the twigs is used to treat diarrhea. Some believe that the plant has value for the treatment of poison ivy.

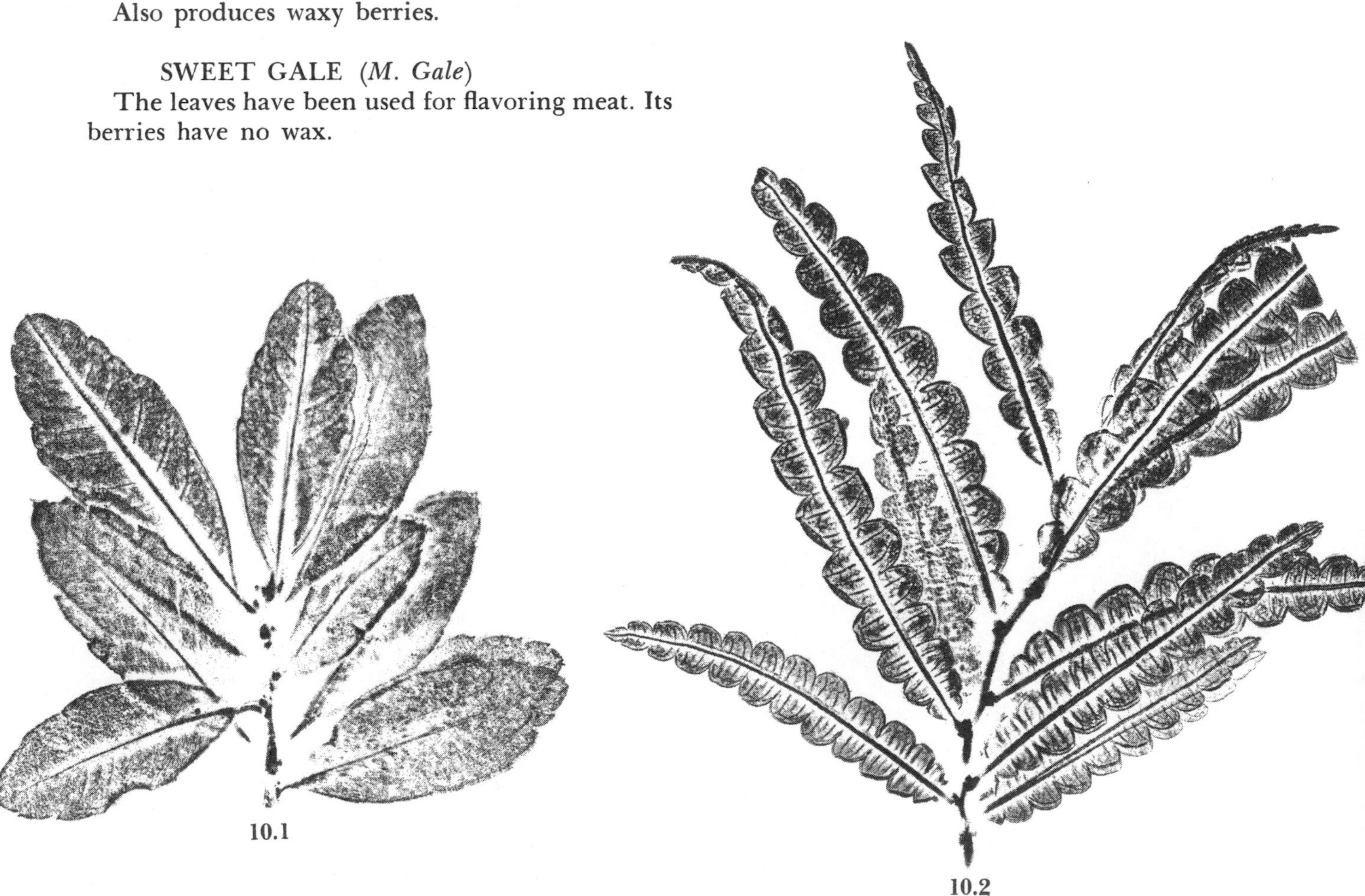

10.1

10.2

JUGLANDACEAE THE WALNUT-HICKORY FAMILY

There are six genera and some 40 species of deciduous trees with alternate, compound, aromatic leaves and edible nuts in this family. Only two of the genera are of interest here: the WALNUTS (*Juglans*) and the HICKORIES (*Carya*). In both, the male flowers are in catkins and the female in small clusters, the pollination being effected by the wind. The fruits of both are popularly called "nuts," but botanically those of the WALNUT are defined as "drupes" while those of the HICKORY are more truly nuts.

There are a number of such obscure differences between the WALNUTS and the HICKORIES, but one practical distinction lies in the fact that the husk is indehiscent in the former and breaks into four valves in the latter. Another dissimilarity is that WALNUTS are widespread around the northern hemisphere while the HICKORIES are almost exclusively American. The wood of many species in both genera is of great value. Incidentally, black walnuts were once made into watch fobs and the hulls of hickories are also good craft materials.

The PECAN and the SHAGBARK are two outstanding HICKORIES. The PECAN (*Carya illinoensis*) (but "Pecahn," with the accent on the ultima to most of us) is a notable southern tree and, with the BLUEBERRY, one of the two important crop trees of American origin. The scientific name indicates the incidental range of this tree as far north as Illinois, but it is in the valleys of Texas and other Gulf States that the wild trees are important. As crop trees they rate highly in Georgia, Alabama, and Mississippi. As might be expected, they rate even more highly in California, where Luther Burbank and others developed the papershell varieties, not to forget Louisiana, where they furnish the distinguishing ingredient of the world-famous pralines and pecan pies. More prosaically, they are used for furniture, flooring, fuel, and for smoking meats.

It might not be amiss to repeat here the contention expressed in the section on SEEDS AND SEED DISPERSAL, that it appears to us that many adaptations can be discerned in the structure of these trees and their fruits. It is worth noting, for instance, that the ridges of the butternut and the reticules of a black walnut help the squirrel to hold the nut, but only after the eating of it has been delayed by the presence of the hull until certain other nuts and seeds are out of season. Likewise, the bark shards of the HICKORIES and various other devices of the nuts themselves serve not so much to prevent squirrels from eating them altogether, but rather to regulate the timing to the best interests of the producing tree. Then some are consumed but enough survive to perpetuate the species.

THE WALNUT GENUS (*Juglans* L.)

A genus of about 15 species of full-size trees that are grown for their nuts and their ornamental value.

11.1 BLACK WALNUT (*Juglans nigra* L.)

This tree is considered by many to be the most valuable in eastern North America. The wood has long been used for furniture and fine cabinetry. During World War I, Boy Scouts had the responsibility of reporting WALNUT trees to the Government so that airplane propellers could be made from the wood. More recently the wood has been used extensively in the production of face veneers and at one time, as the trees became more scarce, the wood was sold by the pound. Recently, poachers have been felling and removing trees while the owner is only temporarily absent!

11.1

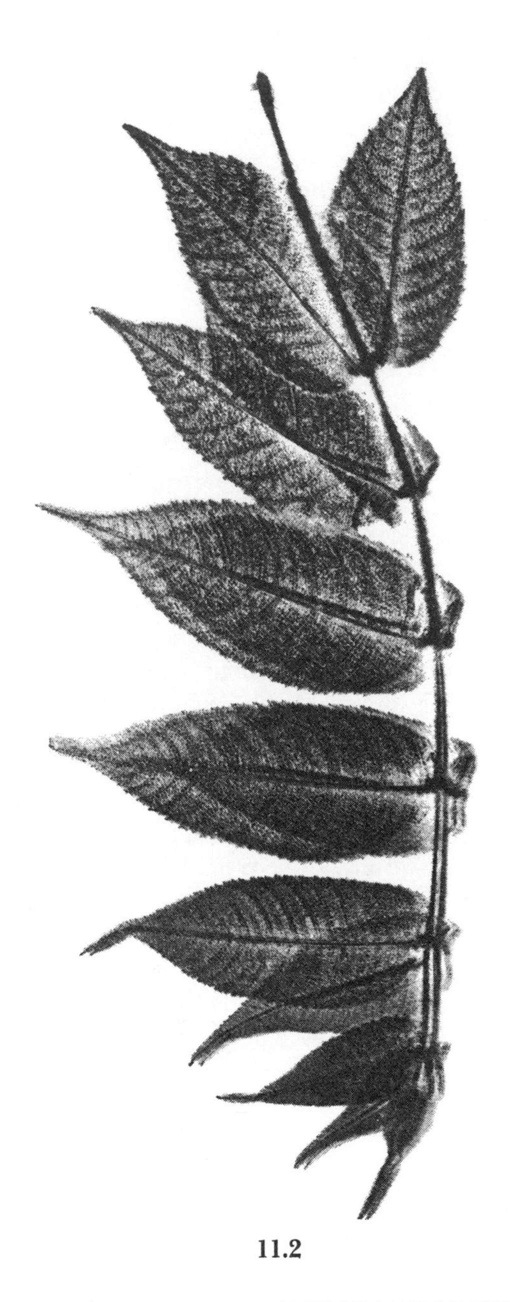

11.2

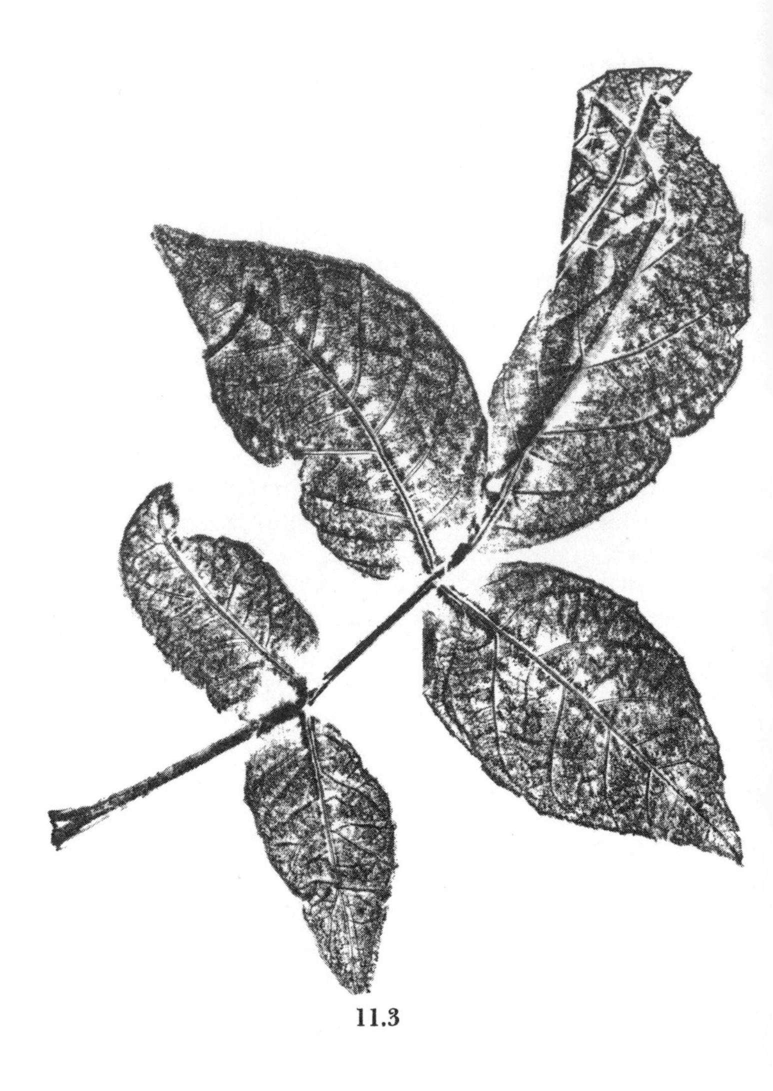

11.3

11.2 BUTTERNUT or WHITE WALNUT
(*J. cinerea* L.)

This is a native American tree but it is local in distribution. The green parts are stickier and more aromatic than BLACK WALNUT. The husks provided the dye used to color the cloth of the Confederate Army uniforms. Sugar may be extracted from the sap. The wood is often used in furniture making. The nuts have a rich, distinctive flavor but the kernels are difficult to remove intact. Oil from the nuts has been used in treating tapeworm infestation and fungus infections.

11.3 ENGLISH or PERSIAN WALNUT
(*Juglans regia* L.)

This important tree is not a native of England, but popular names are not always based on strict accuracy. In the U.S. the crop is grown primarily in California, although the tree grows and produces well throughout the south. There was a large, producing tree at Rockford, Illinois, just south of Wisconsin, and six or eight others that came through the winters there. The leaves are but slightly toothed and the bark is light gray.

THE HICKORY GENUS (*Carya* Nutt.)

There are about 20 species of these strong-wooded, rough-barked trees.

11.4 CHINESE WINGNUT-TREE (*Pterocarya fraxinifolia* or *sp.*)

The print has been checked against Rehder but does not exactly fit his description. It is mentioned in Benson on page 589. Oddly, both the fruit and the leaf rachis are "winged."

11.5 SHAGBARK HICKORY (*Carya ovata* Koch)

Readily distinguished from the other HICKORIES by its 5 leaflets and shaggy bark. The sap can be boiled down to make sugar and the inner bark has been used to give a maple flavor. Withes (see Kephart) are made from the branches and the wood is a favorite for skis and other outdoor-sports supplies. Even billy clubs!

11.5

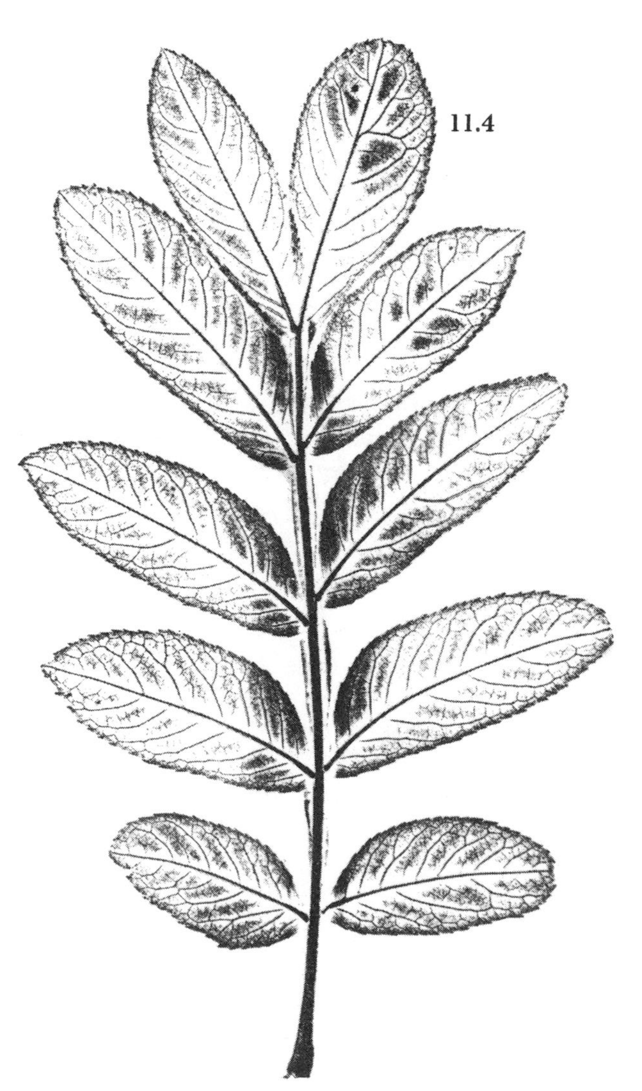

11.4

11.6 MOCKERNUT HICKORY (*C. tomentosa* Nutt.)

This species has 7 to 9 leaflets and is more aromatic than some others. Fernald says that *mocker* comes from the Dutch *moker,* meaning a heavy hammer, but there is considerable subjective mockery in the thickness of the shell of this nut compared with the size of the kernel. With other species, it is widely used for tool handles.

11.6

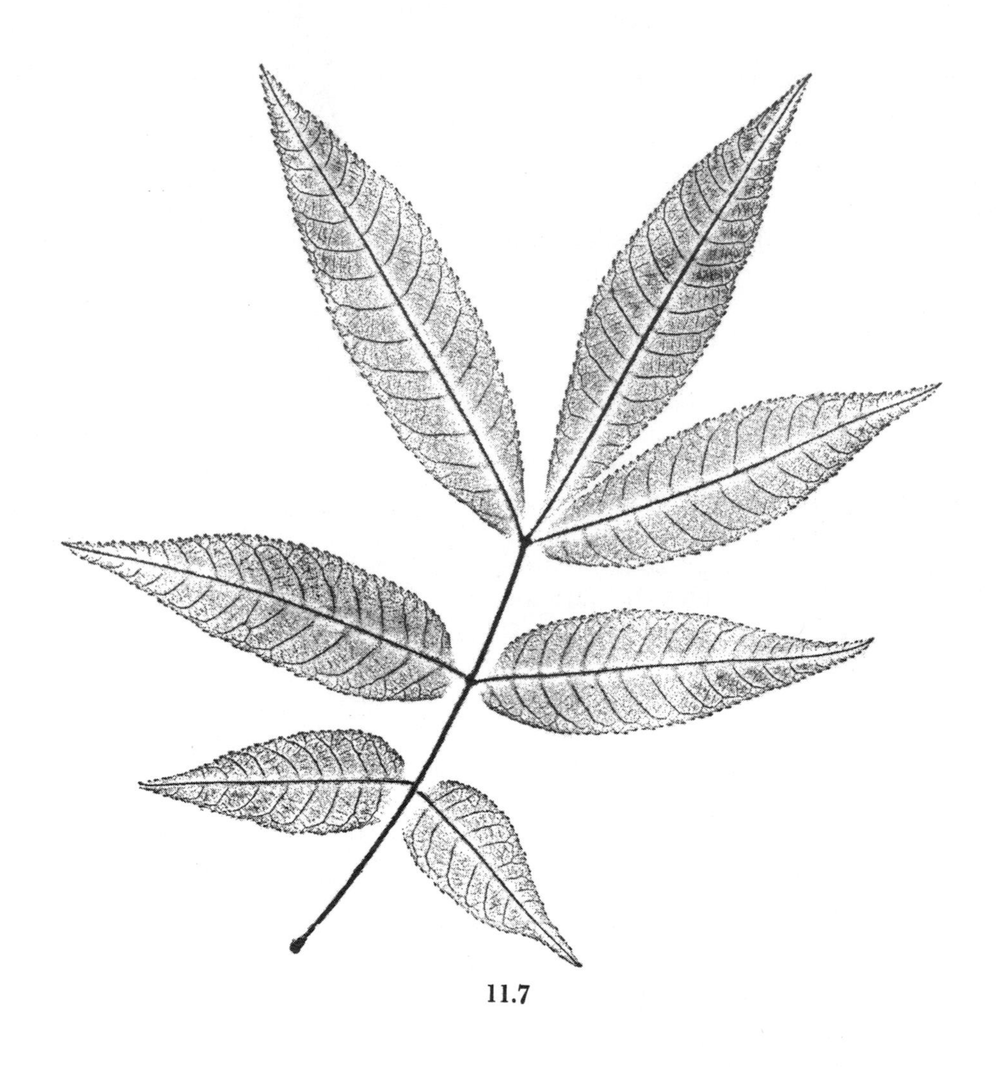

11.7

11.7 BUTTERNUT HICKORY (*C. cordiformis* Koch)

This is a handsome, full-foliaged tree, but it is seldom planted. The elongated, bright yellow buds are a good distinguishing feature. They have a strong, spicy taste and might be tried as a condiment. The nuts are very astringent.

BETULACEAE (CORYLACEAE)

There are over 100 species in the six genera of this family. According to both Bailey and Fernald the family name CORYLACEAE is preferred. However, here we are following Benson, who uses BETULACEAE and then divides the family into two subfamilies, the BIRCHES and ALDERS (BETULOIDEAE) and the HAZELS and HORNBEAMS (CORYLOIDEAE).

The BIRCHES are perhaps the most outstanding genus in the group, but several others are also interesting. The HAZELS supply food for man and animals. The BIRCHES, HORNBEAMS, and HOP-HORNBEAMS have their seed-like fruits winged, inflated, or otherwise disposed toward dispersal by the wind. The ALDERS are in transition, with a tendency to forgo the wings in favor of dissemination by floating on water.

The leaves of most BIRCHES turn yellow in the fall and this might be interpreted to imply that a signal is being given to the mice and other rodents to supplement the wind in the dispersal of the seeds. We could not base such an inference on one genus alone, but the leaves turn yellow on practically all trees that have winged seeds, for instance, CATALPA, TULIP-TREE, LINDENS, ELMS, and AILANTHUS. There are exceptions.

BIRCHES and ASPENS are important in the north as nursetrees of evergreen conifers. The conifer seedlings cannot withstand full sunlight in their early stages and they are shielded from it by the broad-leaved trees, which they ultimately supplant in the community succession.

THE BIRCH GENUS (*Betula* L.)

A genus of about 40 species of smooth-barked trees and shrubs. Along with the other members of the family, the BIRCHES bear staminate and pistillate catkins. The seeds (fruits) are small, winged, and wind-borne. The wood is hard, heavy, strong, and tough. It often has a beautifully figured grain that will accept a high polish. YELLOW and CHERRY BIRCH are especially valued for making furniture.

12.1 TRIANGLE or GRAY BIRCH (*Betula populifolia* Marsh.)

This native birch already has several names but none are satisfactory. It sports three sorts of triangles, 1) the leaf shape, 2) the outline of the clump, and 3) a black "eyebrow," where each large branch joins (or did join) the trunk. This is the usual "CLUMP-BIRCH" of the nurseries, but it is short-lived. Homeowners who want a BIRCH that will endure should order the CANOE BIRCH if they live within its range.

12.1

12.2 CANOE or PAPER BIRCH (*B. papyrifera* Marsh.)

The birchbark canoe has been said to be the greatest contribution of the Indian to civilization, which doesn't seem very flattering. The wood of this tree is a superior fuel and is also favored in modern kitchen cabinetry. The bark is the most

12.2

12.3

celebrated part of the tree. Bernard Mason tells how to obtain it from commercial sources instead of desecrating live trees. His chapter on this subject is one of the best. The bark will burn, wet or dry.

12.3 CHERRY or SWEET BIRCH (*Betula lenta* L.)

The dark, glossy bark is distinctive. The crushed leaves have a wintergreen odor and these and other parts are medicinal. The sweet sap can be boiled down for sugar or may be used to provide vinegar. The twigs in past days were often used for making besoms or birchbrooms. In some parts of the country root beer is called "birch beer."

12.4 RIVER BIRCH (*B. nigra*)

The most southerly of the BIRCHES. There is a legend that a detachment of fleeing Confederate soldiers left a trail by stripping and eating the inner bark of this tree. Usually the hope of subsisting on tree bark is a cruel hoax. RIVER BIRCH lines river banks most attractively in some places and its light brown, tattered bark makes it distinctive in cultivation. Both the bark and the wood are good fuels and the bark, especially, is valuable for craft work.

12.4

YELLOW BIRCH (*B. lutea* Michx.)
Usually, but not always, weeping. The most popular form is the native species with leaves larger than those of SWEET BIRCH and bark intermediate between that of SWEET and the RIVER BIRCH. It is faintly wintergreen scented.

EUROPEAN BIRCH (*B. pendula* Roth)
Also usually, but not always, weeping in form. The most popular variety is the CUTLEAF WEEPING BIRCH (*B. p.* var. *dalecarlica* Rehd.) named for Dalecarlia or Delarne in Sweden. The leaves are similar to those of GRAY BIRCH but much smaller and deeply incised. The bark of the EUROPEAN BIRCH is used for tanning Russian leather. It is also famous for its incorruptibility in contact with soil bacteria. Perhaps it contains a bactericide.

12.5

THE ALDER GENUS (*Alnus* B. Ehrh.)

About 30 species of shrubs in the eastern states and trees in the western states and Old World. The leaves are slightly lopsided and conspicuously veiny. The flowers are wind-pollinated. The seeds show evolutionary transition correlated with the wet-or-dry habitat of these trees; some species are winged and some are not. These were among the first plants to be shown living in biotic relationship with fungi.

12.5 SMOOTH ALDER (*Alnus serrulata* Willd.)
The ALDERS are noted for growing in wet ground and for ranging far northward, where they are used by woodsmen for snares and many other woodcraft articles. The bark will give a brown dye and when chewed is used to treat wounds and ulcers.

12.6 SPECKLED ALDER (*A. rugosa* Spreng.)
This species is recognized by the tooth-lobed leaves and the conspicuous lenticels on the twigs.
RED ALDER (*A. rubra*) is a leading hardwood in the Pacific Northwest. It is used for furniture.

12.6

EUROPEAN ALDER (*A. glutinosa* L. Gaertn.) is a tall, usually narrow tree. There is a spreading specimen in the Morton Arboretum, but the ones to be seen in Cincinnati are slender. The leaves are sticky when young and are indented at the tip, rather than pointed. The wood is noted for resistance to decay when permanently submerged.

THE HAZEL GENUS (*Corylus* L.)

The HAZEL genus contains some 15 species of nut-bearing trees and shrubs, some native to the U.S. and others in Europe. They are common in cool temperate regions of both hemispheres. The leaves of our native species turn red, yellow, and other colors in the fall, thus attracting a variety of disseminators. The EUROPEAN HAZEL or FILBERT (or COB) is said to have been named after one St. Philibert. The distinction between filberts and cobs has been obscured by commercial breeding practices, although originally cobs were from *C. Avellana* and filberts from *C. maxima*. The trees may also be found as ornamentals, sometimes in a purple-garbed variety.

12.7 AMERICAN HAZEL or HAZELNUT (*Corylus americana* Marsh.)

Produces a nutritious and palatable nut, but is of no commercial importance.

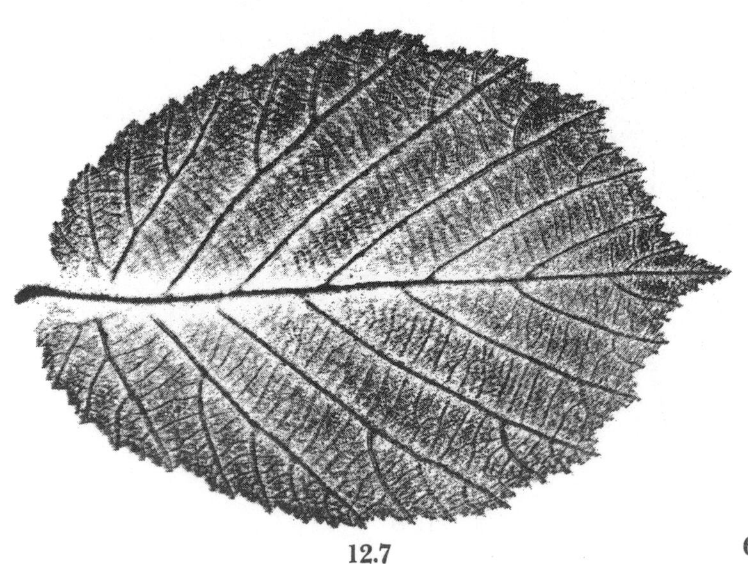

12.7

12.8

12.8 EUROPEAN HAZEL or COB (*C. Avellana* L.)

A prominent forest tree of Europe that is cultivated for the nuts in Southern Europe, England, and Oregon.

THE HORNBEAM GENUS (*Carpinus* L.)

These are small trees with rather smooth gray bark, the trunk and larger branches appearing "muscular." The "fruit" is of an unusual structure, with a three-lobed bract and a small "seed." They are probably disseminated in part by wind and in part by rodents. The leaves turn red in the fall; the wood is very hard. Some of the European species of this genus, which are widely planted, can be recognized by the thicker texture, depressed venation, and the heart-shaped base of the leaf.

12.9 AMERICAN HORNBEAM or BLUE-BEECH (*Carpinus caroliniana* (Walt.)

The leaves are smaller than in the HOP-HORNBEAM and turn red; the trunk is muscular and the bark gray. The seeds are attached outside the bract.

12.10 EUROPEAN HORNBEAM (*C. Betulus* L.)

THE HOP-HORNBEAM GENUS (*Ostrya* Scop.)

This genus is known to Americans almost entirely from our native species. The European cognate (*O. carpinifolia*) is rarely seen outside of botanic gardens. The "seeds" are inclosed within an oblong bladder, which aids variously in seed dispersal and is good craft material. The leaves of our native species turn a rich yellow.

N.B. Trees of both genera are commonly called IRONWOOD, which is unfortunate. It does not improve matters much to apply the name FALSE-IRONWOOD to the genus *Carpinus*.

12.11 AMERICAN HOP-HORNBEAM (*Ostrya virginiana* Koch)

The leaves are larger and more streamlined; the trunk is strictly cylindrical, the bark brown with close-set vertical fissures; and the seeds are in bladdery wrappings. It is a slow-growing shade tree.

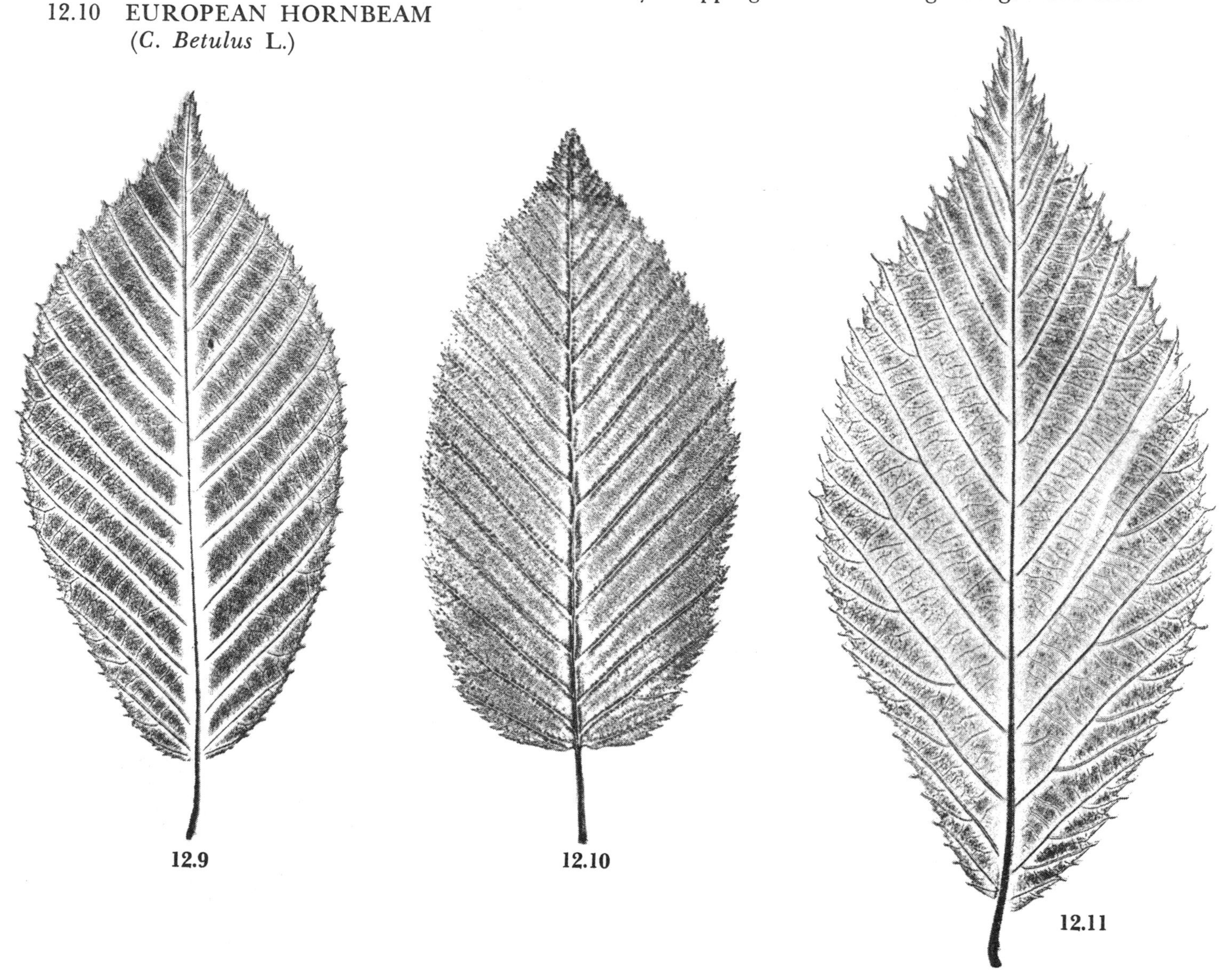

12.9 12.10 12.11

The family is best known for the OAKS, although the BEECHES and CHESTNUTS are also important. In addition, it includes the South Temperate genus *Nothofagus* and the two Pacific Coast genera *Lithocarpus* and *Castanopsis,* for a total of six genera and about 600 species. The flowers are in catkins and are typically pollinated by the wind but, in the CHESTNUTS, largely by insects.

The theory is fairly well substantiated that the reason for so many more species of OAKS in America than in Europe is that our mountains run north and south. It is thought that the Mediterranean Sea and the mountains around it became a wall that the southward-retreating trees could not pass to escape the rigors of the climate when glaciers moved southward in both hemispheres.

The fruits in this family are various types of nuts; they provide food for wildlife and humans and disseminating devices for the trees. Some acorns and beechnuts, and all chestnuts, are protected from squirrels in cups or burs, probably for reasons similar to those discussed in the Walnut-Hickory family. The leaves of many OAKS turn red in the fall, which I interpret to be a device to signal birds to aid in the dissemination of the seeds. This whole problem needs study, observation, and controlled experiment, since we know almost nothing about the adaptational value of leaf color.

A DISTINCTION BETWEEN BITTER AND ASTRINGENT

Acorns and coffee are both bitter and astringent, but we often confuse the two, and astringency or puckeriness may be miscalled bitterness. Bitterness is a true taste (bite a LILAC, TEASEL, or CATALPA leaf), but astringency is a physical sensation caused by tannins. To get a pure sensation of astringency bite a BLACKBERRY, AGRIMONY, or STRAWBERRY leaf.

THE BEECH GENUS (*Fagus* L.)

The Beeches comprise about 10 species of ornamental trees with smooth, light-gray, "elephant-hide" bark. These deciduous trees favor the cooler portions of the northern hemisphere. The ovoid nuts are sharply 3-angled and occur singly or in pairs within a prickly covering that breaks open in four parts (valves).

13.1 AMERICAN BEECH (*Fagus grandifolia* Ehrh.)

An important forest tree, especially in OAK-BEECH-HEMLOCK climax communities, and beautiful lawn specimens, but seldom planted. Its leaves were once gathered before the frost and used for stuffing mattresses. Destructive distillation of the wood yields creosote, which is an important drug and preservative. The nut crop is not dependably annual, but is of great importance to wildlife. Kephart says that the oil compares favorably for cooking with the best olive oil, an item of more historical than present-day importance. Some believe that lightning does not strike Beech trees.

13.1

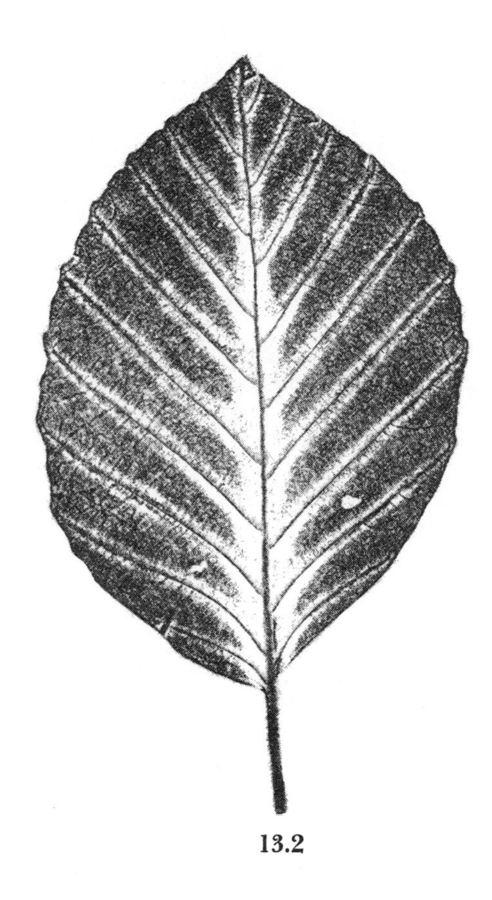

13.2

13.2 EUROPEAN BEECH (*F. sylvatica* L.)
A tree of extraordinary beauty and *the* BEECH of cultivation. Easily distinguished from our species by its darker bark and fringed leaves. The wood has many uses in England, as attested by Edlin. The seeds are freely produced in England (not here) and the trees may be underbedded with seedlings. This tree varies greatly, and copper-leaved, cut-leaved, weeping, and other forms are common in cultivation.

THE OAK GENUS (*Quercus* L.)

The OAKS in general and the WHITE and PIN OAKS in particular are surely among, if not *the*, favorite trees of Americans. We appreciate them both growing wild in nature and planted on properties.

There are over 300 species of deciduous or evergreen trees from the temperate regions of the northern hemisphere and the mountains of the tropics in this genus. The leaves are borne on short petioles. The fruit is a one-seeded nut (acorn), more or less inclosed in a cup composed of many bracts partly or completely united. This is perhaps the only American tree with a "fruit" name apparently different from the tree name, but actually it is not, since "acorn" comes from sources close to the German *Eich Kern* (oakseed).

The OAKS are pollinated by the wind and disseminated chiefly by squirrels. They depend largely on abundance of cropping, with enough nuts not eaten to insure race continuity.

Our American OAKS are divided into "WHITE," with round-lobed leaves and the acorns taking one year to ripen, and "BLACK," with the tip and side lobes (if any) ending in bristles. In these, the acorns take two years to ripen. Such a classification, based on a limited area, does not account for intermediate forms found in other parts of the world.

The OAKS have been particularly noted for the quality of their timber, which has been made into everything from treenails to ship masts. For the properties and uses of ENGLISH OAK wood see Edlin. For variation in the qualities of American OAK woods, see the younger Hough and Schoonover.

Both the bark (especially of the BLACK OAKS) and the insect-initiated galls are rich in tannin and have been used for ages both medicinally and for tanning leather. The Old World ALEPPO OAK (*Quercus infectoria*) was the original source of the commercial galls. The specific epithet comes from Latin and means "to dye" rather than "to infect." Both cured oak leaves and the galls are a source of food for grouse.

The LIVE OAK (*Quercus virginiana*) of the Deep South is an outstanding tree for both beauty and utility. The large and abundant acorns are indispensable to many forms of wildlife and can be eaten by man when properly prepared.

The TURKISH OAK (*Quercus Cerris*) (not the native Turkey Oak) is a timber tree of the Old World and an ornamental in this country. The acorns are smaller than those of the BUR OAK, but the cup fringes are even longer, making this craft material of more than usual appeal.

THE WHITE OAKS

13.3 WHITE OAK (*Quercus alba* L.)

This is an important upland tree, the wood long of special value and the tree much appreciated for its robust symmetry and red autumn color. The wood is among the kinds charcoaled for cookouts. It is made into barrels because there is no chemical in it that taints liquid contents. The acorns are relatively sweet and can be eaten if ground, leached, and cooked. The bark and some of the galls of many OAKS are rich in tannin used medicinally and for tanning leather. The galls also serve as food for grouse.

13.4 BUR or MOSSYCUP OAK (*Quercus macrocarpa* Michx.)

The BUR OAK is a tree of wide range and extends far northward into Canada—the farther north, the smaller the acorns. In the Rock River valley of northern Illinois they are tiny; in the Wabash valley at the south end of the same state they are enormous. They are very good for craft work and edible when properly prepared. The corky twigs are also ornamental. The value of the cork to the tree itself has been discussed in the introductory section on Tree Bark.

13.5 OVERCUP OAK (*Quercus lyrata* Walt.)

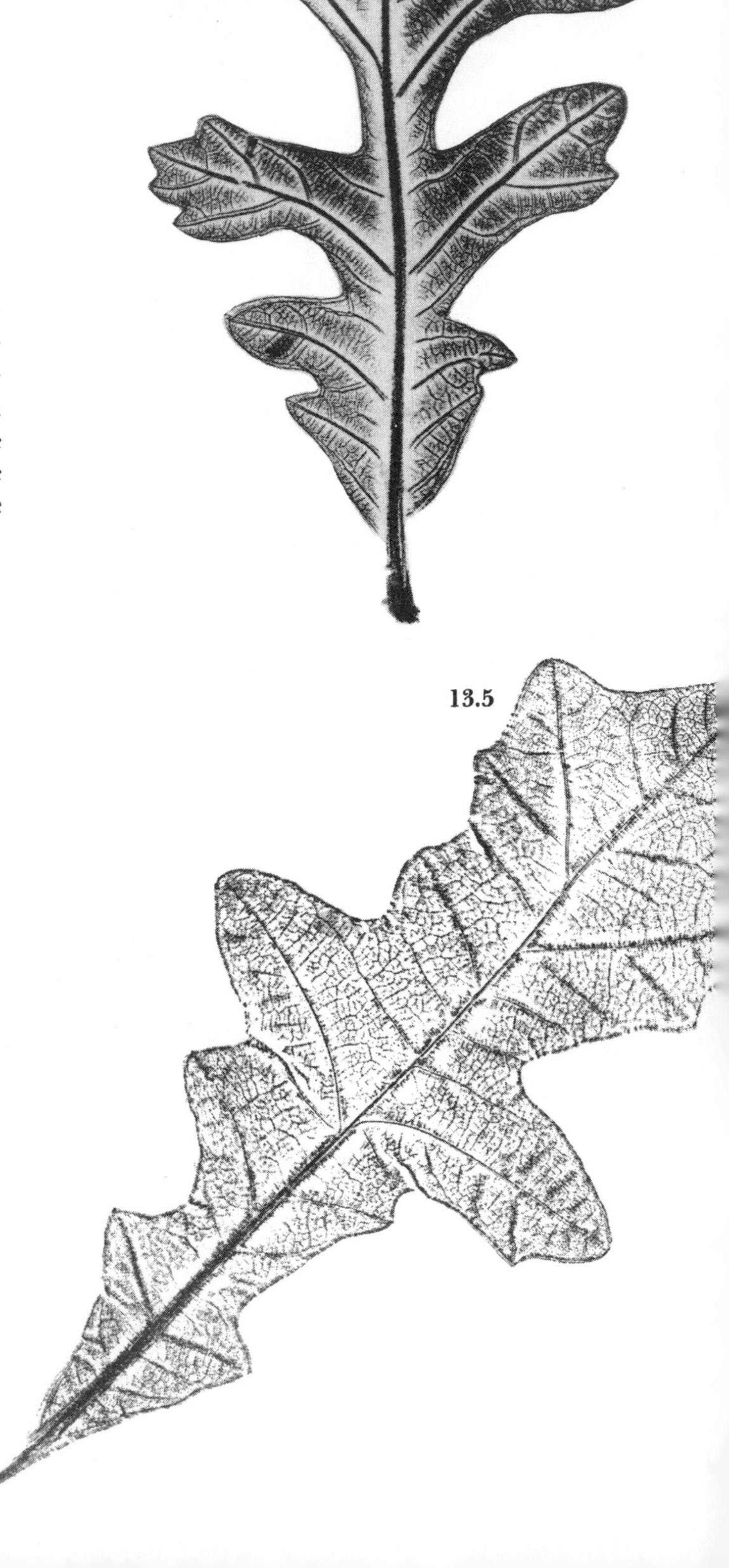

13.4

13.5

13.3

13.6

13.8a CHESTNUT-OAK; ROCK-CHESTNUT-OAK (*Quercus Prinus* L.) (Fernald, not Bailey)

Tree with deeply furrowed bark growing on rocky outcrops.

13.8a

13.6 ENGLISH OAK (*Q. Robur* L.)

13.7 SWAMP WHITE OAK (*Q. bicolor* Willd.)

13.7

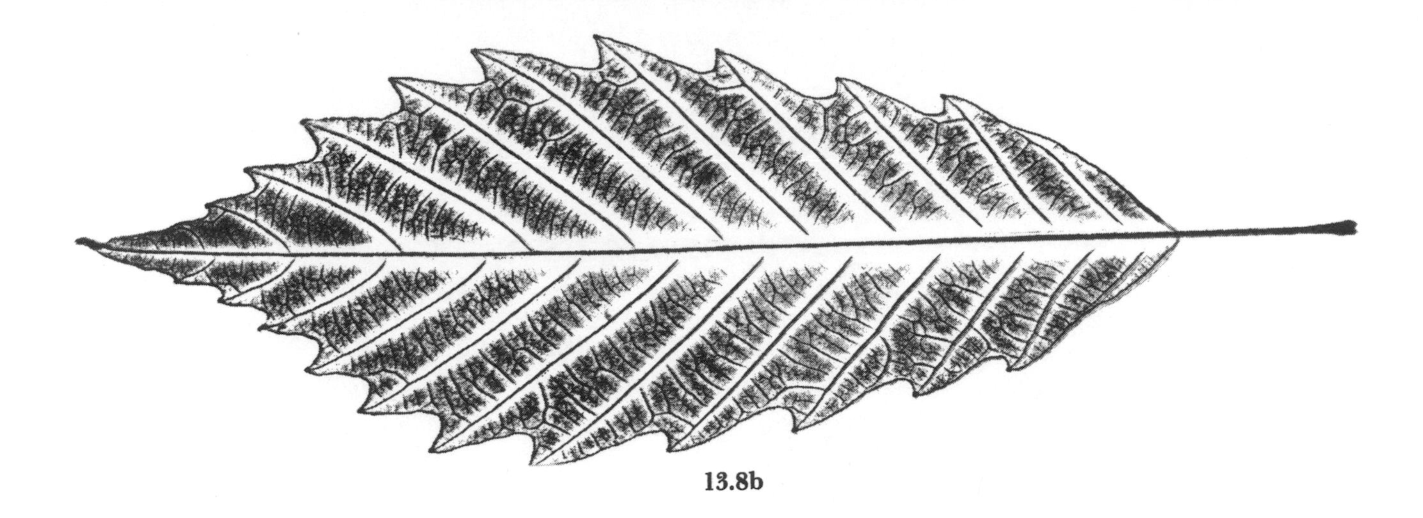

13.8b

13.8b CHESTNUT-OAK; YELLOW OAK
(*Q. Muehlenbergii* Engelm.)

Bark thin, becoming flaky when mature. Found on dry calcareous slopes and ridges, but also in rich bottomlands.

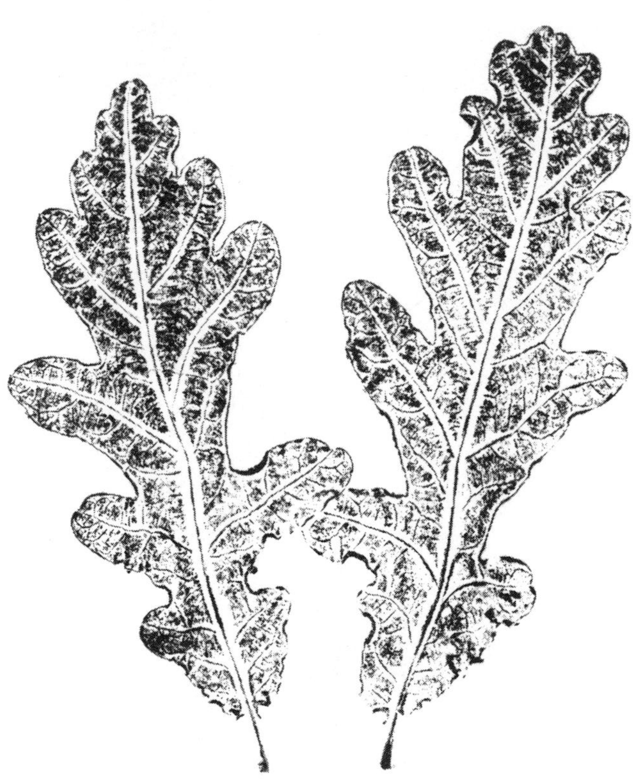

This leaf-print is of an interesting hybrid of *Quercus Cerris* and *Q. Robur*. The tree is located in the vicinity of Muhlenberg College, Allentown, Pa.

THE BLACK OAKS

13.9 PIN OAK (*Quercus palustris* Muenchh.)

By far the most planted of the hardy OAKS, partly because it is truly beautiful and partly because, being a swamp tree, it has no tap root and is easy for nurserymen to transplant. The small acorns are pied or two-tone. The leaves turn red. As with the PINES and FIRS, this tree starts out in pyramidal form, highly competitive at the ground level. When the tree starts to mature, the top spreads wide—likewise competitive under forest conditions.

13.9

13.10

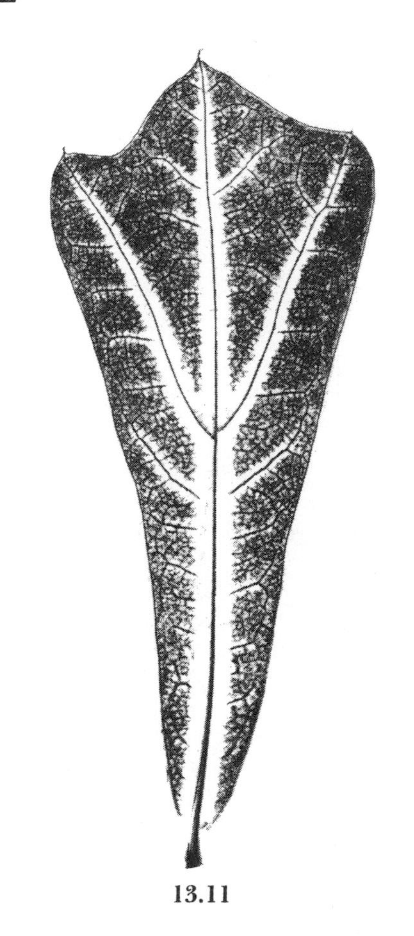

13.11

13.10 CORK OAK (*Quercus Suber* L.)

A native of the Old World but it thrives in California, the Gulf States, and cool greenhouses.

About every eight years the layers of bark are removed from the trunk and large branches. The cork is boiled to remove certain chemical impurities and then it is shipped to processing plants.

13.11 WATER-OAK (*Quercus nigra* L.)

13.12 BLACK or QUERCITRIN OAK (*Quercus velutina* Lam.)

The two trees that most attracted the first American colonists were this and SASSAFRAS. The pioneers found that the inner bark of this OAK gives a bright and permanent yellow dye and that as with all Oaks (but especially BLACK OAKS) it is unsurpassed for tanning. QUERCITRON is the crude material and QUERCITRIN the purified derivative. The acorn of this species is yellow and very bitter and astringent.

13.12

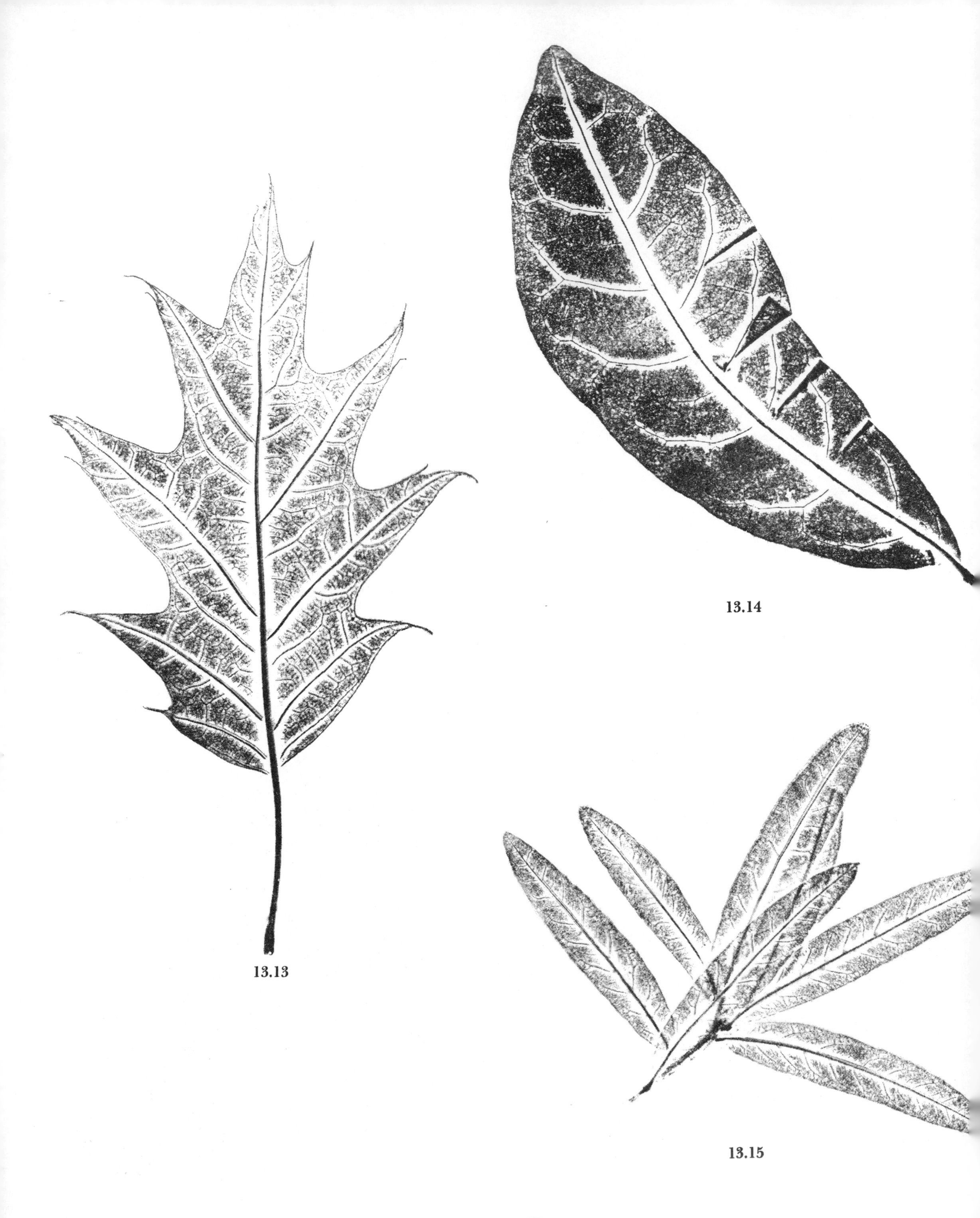

13.13

13.14

13.15

13.13 RED OAK (*Quercus rubra* L.)

The variation in the form of BLACK OAK leaves, especially on saplings, may be so great as to overlap with the typical form of the RED OAK leaves, thus making identification difficult. However, the cup of the acorn can be a distinguishing factor, since in the RED OAK it is a more shallow, saucer-like structure than the deeper, cup-like form found in the BLACK OAK. (Black=cup; Red=saucer)

13.14 SHINGLE OAK (*Quercus imbricaria* Michx.)

13.15 WILLOW OAK (*Q. Phellos* L.)

13.16 KOREAN or SAWTOOTH OAK (*Q. acutissima*)

13.17 SCARLET OAK (*Q. coccinea* Muenchh.)

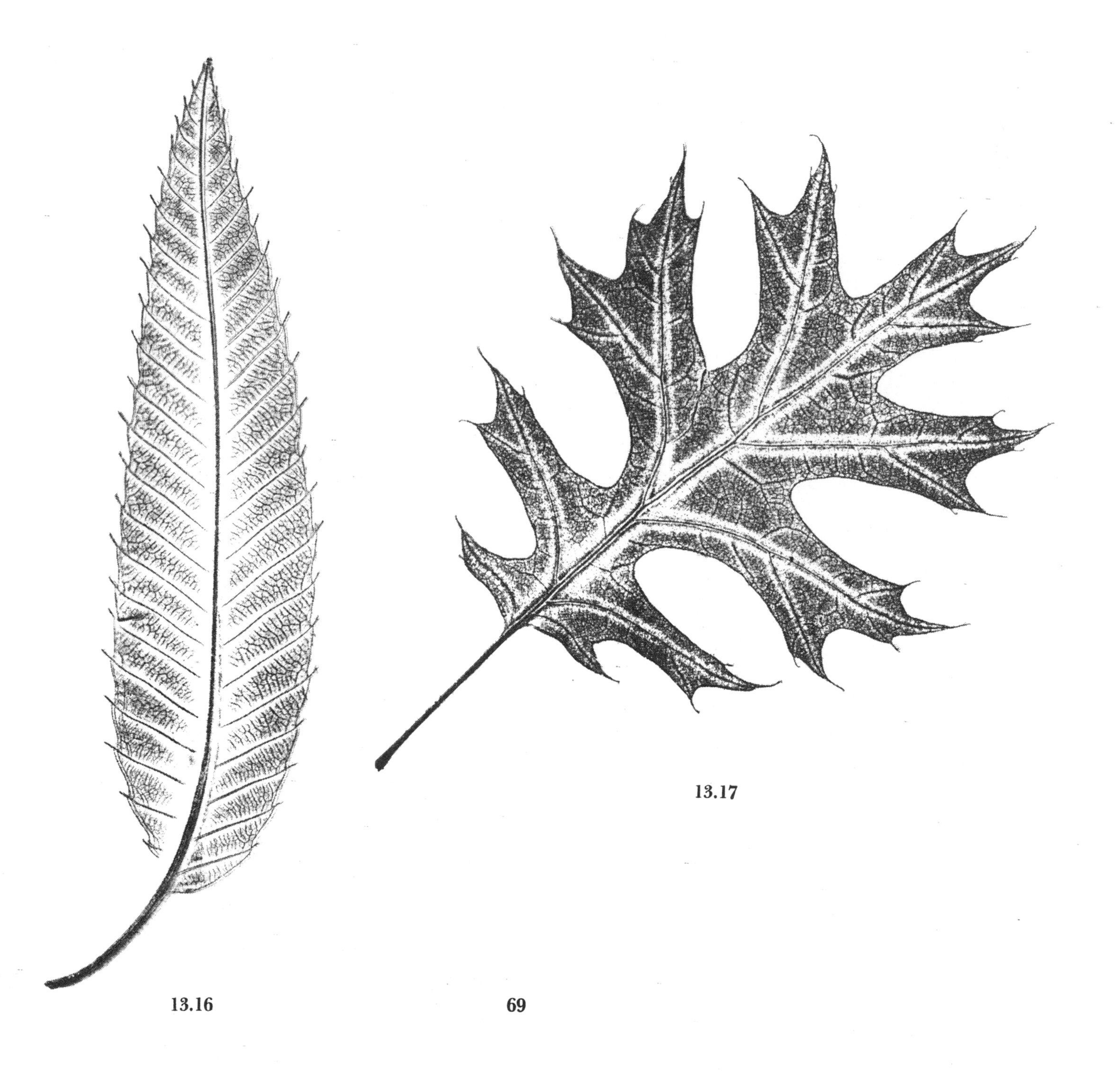

13.17

13.16

THE CHESTNUT GENUS (*Castanea* Mill.)

Some 10 species of deciduous trees and shrubs are included in the genus. The bark is furrowed and the branchlets are without terminal buds. The AMERICAN CHESTNUT trees are now mostly gone from their former haunts as a result of the Chestnut Blight (1904, N.Y.C.). Young saplings can still be found arising from the old roots, but after a few years they, too, succumb. Only occasionally do they become old enough to bear a few nuts before they die back.

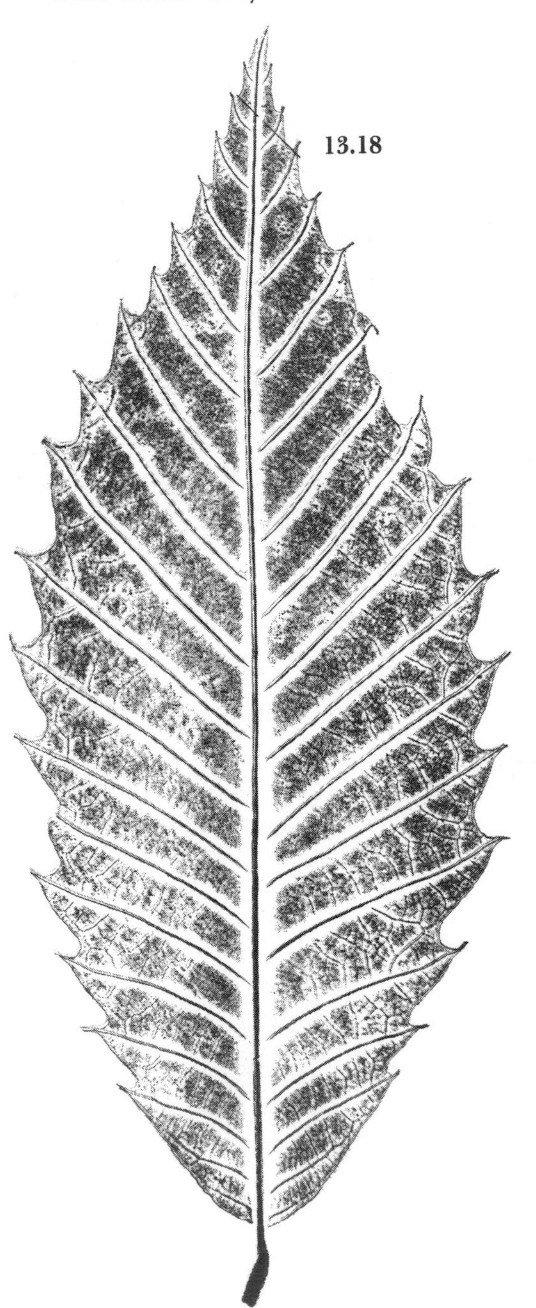

13.18

Although the big spiny burs have received far more attention, the most notable feature of the CHESTNUTS is their insect-pollinated flowers. These are in catkins, as with the rest of the family, and only the stamens of the staminate flowers are in evidence; *but* they are whitish rather than green, and they are fragrant. When in bloom the SPANISH and AMERICAN CHESTNUTS seem decked in white, and are visible for a long distance. The spiny burs protect the unripe nuts from the squirrels. As the nuts ripen they fall free from the bur; some are eaten, the rest carry on the race.

AMERICAN CHESTNUT wood was once important for lumber and millwork and as a source of tannin.

For the story of how this glorious tree was seen before it was doomed by the Old World fungus to which it was not resistant, see both the elder Hough and the younger Hough listed in the bibliography.

13.18 AMERICAN CHESTNUT (*Castanea dentata* Borkh.)

13.19 SPANISH or ITALIAN CHESTNUT (*C. sativa* Mill.)

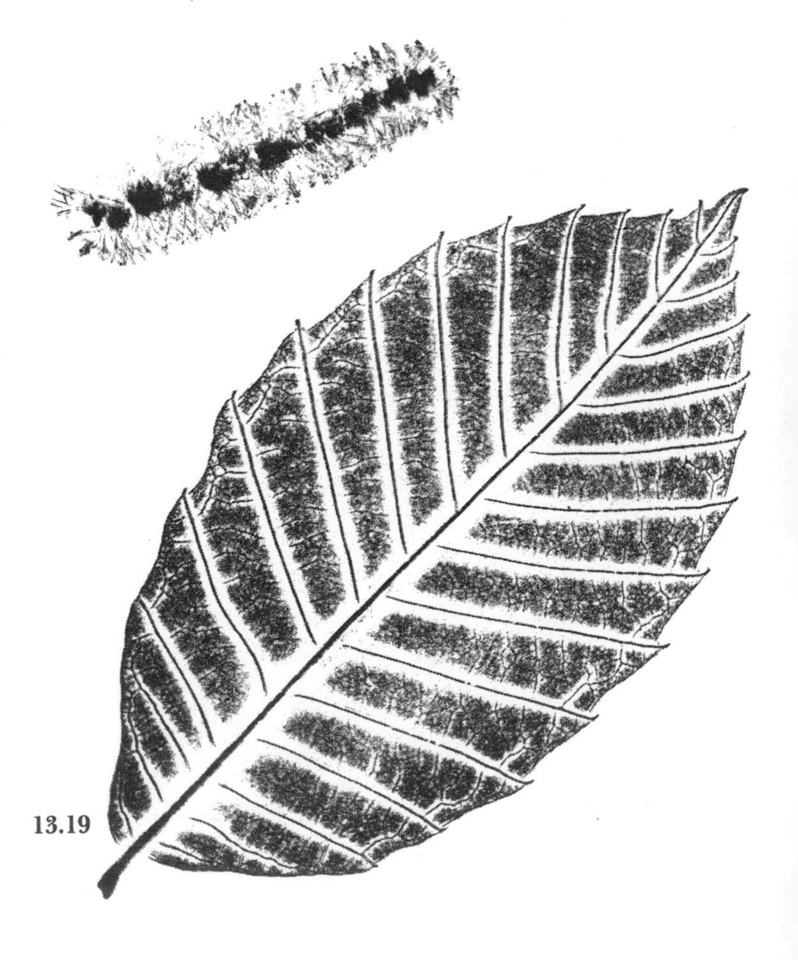

13.19

The family consists of 15 genera and more than 150 species of deciduous trees and shrubs, many of which are important street and shade trees. The leaves are alternate, simple, serrate, and pinnately veined.

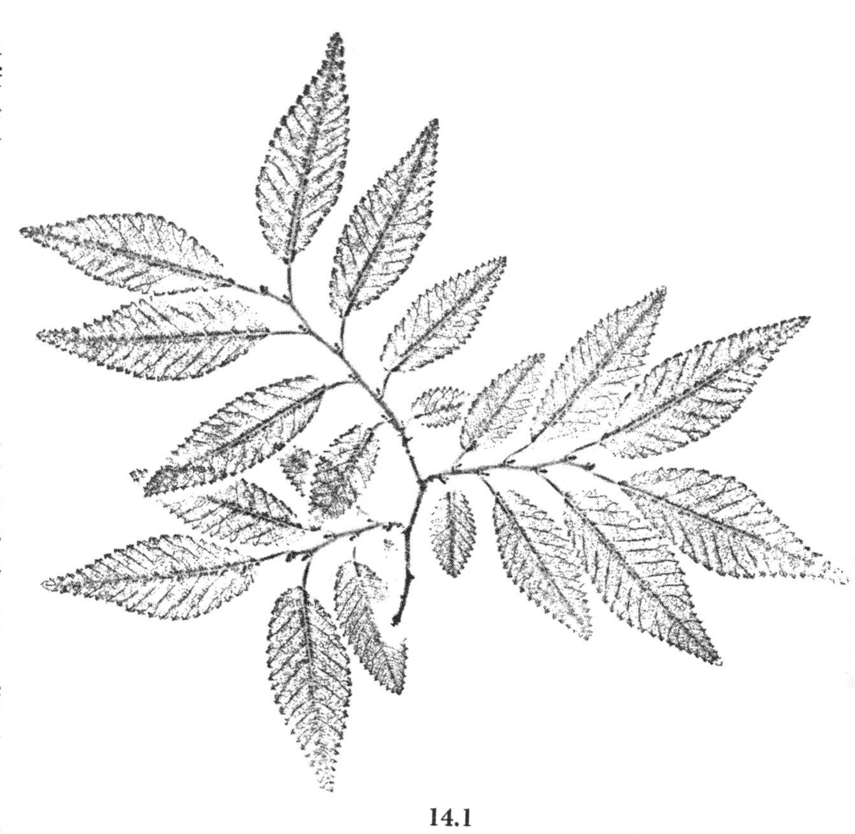

14.1

THE ELM GENUS (*Ulmus* L.)

The genus contains about 18 species native in the northern hemisphere. Their leaves are either singly or doubly serrate and are unequal at the base. The wood is very resilient and tough and was used for wagon-wheel hubs and for other items with a similar need for strength. Elm branches will bend far over without breaking, so that it is a very safe tree for climbing. The bark is corky and non-abrasive. When cut with a pen-knife, the bark of the AMERICAN ELM shows alternate light and dark chocolate-colored layers; SLIPPERY ELM does not.

Susceptible to the Dutch Elm disease, which is transmitted by several elm-bark beetles, many stately old elm trees of New England and other parts of the country have now died. The twig-feeding beetles inoculate healthy trees with the fungus *Ceratostomella ulmi,* which then clogs the water-carrying cells of the tree and kills it.

14.1 BRITTLELEAF ELM (*Ulmus parvifolia* Jacq.)

Only planted. The bark is plate-like. Leaves break when folded. Resists Dutch Elm disease.

14.2 SLIPPERY or RED ELM (*Ulmus fulva* Michx.)

A common and widespread native tree ranging far northward. The inner bark, an emergency or survival food, also supplies a strong fiber and is the part used medicinally for soothing inflamed tissues. The notorious "spitball" curve of baseball, now outlawed, was originally dependent upon a touch of this mucilage surreptitiously placed on the ball. Children once chewed the bark as a "chewing gum." The leaves feel sticky if pressed between the fingers. They and the bark are fragrant when dry. The leaves turn black under the tree, leading to easy identification.

14.2

14.3 AMERICAN ELM (*Ulmus americana* L.)
A large spreading tree of the eastern half of the U.S. and Canada and of particular distinction in New England. The wood has been used in furniture, containers, and in supplies for the apiary, dairy, and poultry industries.

14.3

14.4

14.4 ENGLISH ELM (*Ulmus procera* Salisb.)
A tree with one oak-like trunk, which also sends up sprouts from the roots.

14.5 CHINESE, or better, SIBERIAN ELM (*U. pumila* L.)
The leaves are relatively small. It often runs wild.

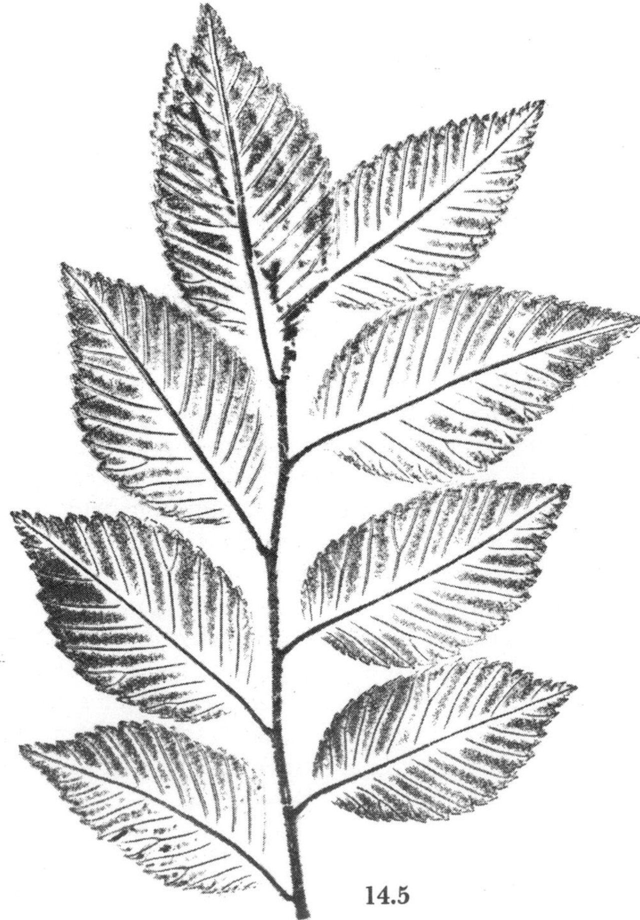

14.5

THE HACKBERRY GENUS (*Celtis* L.)

There are about 70 species of trees and shrubs in this genus, of which most are deciduous but a few are evergreen. The leaves are 3-nerved at the base. The fruit is a drupe with a thin pulp and a thick outer coat.

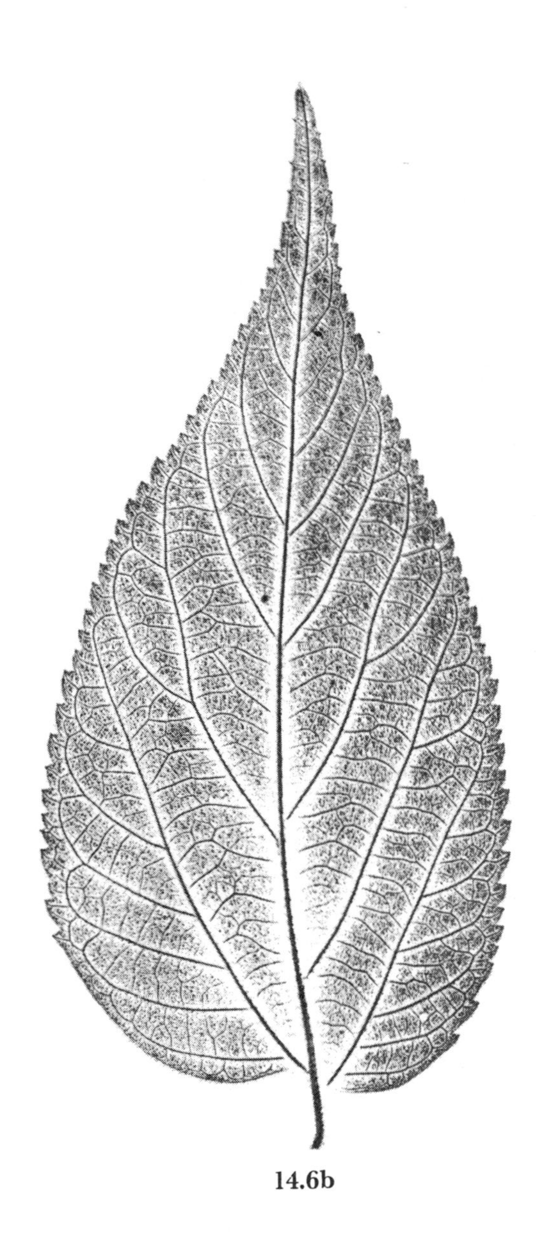

14.6b

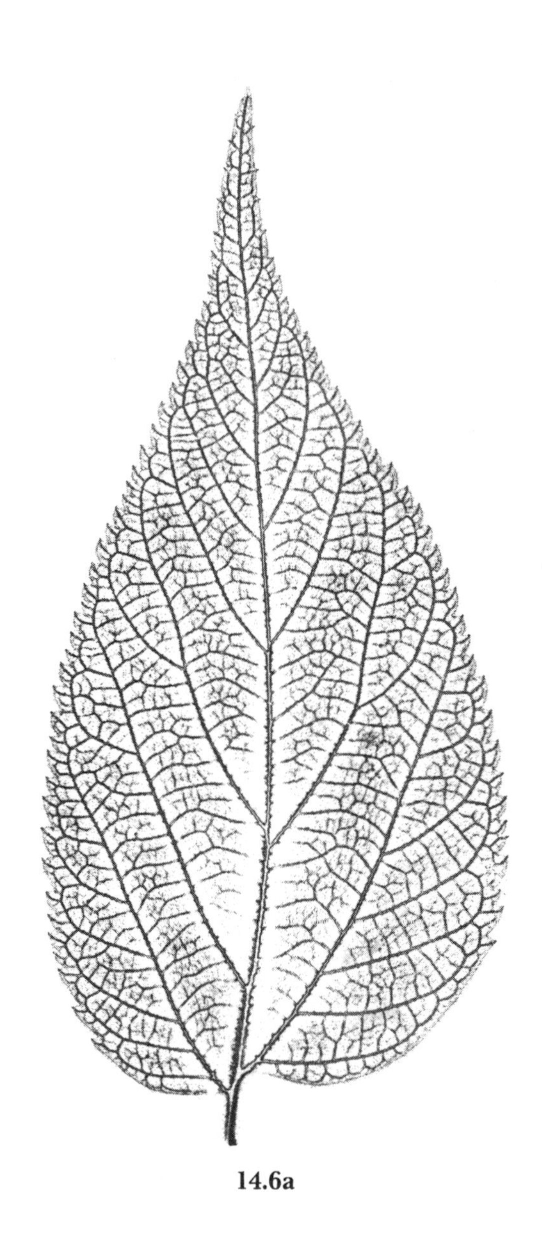

14.6a

14.6 HARDY HACKBERRY TREE (*Celtis occidentalis* L.)

An elm type of tree easily recognized by its "witches' brooms" and by the larger branches, which turn abruptly upward. The sweet berries are important wildlife food and edible to humans. A strong sugar can be made from them. The seeds are said to pop like popcorn if overheated. The bark varies greatly in appearance as the tree ages and can be used in craft work, topographic maps, and window displays. The very rough leaves can be used in camps for cleaning pots and pans. According to the elder Hough, the wood is very hard. Hackberry is occasionally planted in cities.

MORACEAE

About 55 genera and possibly 1,000 species of trees, shrubs, herbs, and some vines are included in this family. Many of them produce a milky juice.

The characteristics of the family are best exemplified by the Mulberry genus, in that they have edible "berries" (multiple accessory fruits) compounded from adjacent carpels and the pollination of many separate flowers. They have been in cultivation for over 4,000 years.

The WHITE MULBERRY has grayish-yellow bark, which may aid in attracting disseminating birds and mammals to the fruit. The pinkish bark of the PAPER-MULBERRY may serve likewise. The leaves of the WHITE MULBERRY are, or were, a principal food of the silkworm, and the first attempts to make artificial silk were made with its wood. Incidentally, it was during an attempt to cross-breed silkworms with gypsy moths that some of the latter escaped and became the progenitors of the scourge which annually attacks the forests of the northeast.

While the RED MULBERRY is still to be found, its seeds do not germinate readily in hard open soil and it is losing ground to the introduced WHITE. The berries are favorites with birds, so MULBERRIES are sometimes planted near CHERRY-TREES as a counter-lure. The BLACK MULBERRY (*Morus nigra*) provides the best fruit, much eaten in some parts of the world and valuable also as food for hogs and poultry.

THE MULBERRY GENUS (*Morus* L.)

A very variable genus for which many species have been described. A conservative estimate, however, would be about 12 species. The fruits are structured like a blackberry, as described in the introductory section on Fruits and Seeds.

15.1 WHITE MULBERRY (*Morus alba* L.)

A broad, smallish tree with glossy and rather small leaves. Its leaves are the chief food of the commercial silkworm and, on the theory that there was something "fibrogenetic" about the tree, the first experiments on the artificial synthesis of rayon utilized its wood as the raw material. Selected strains with more luscious berries are commercialized. The inner bark fibers of the trunk are very strong and can be used for rope-making. The bark of the roots provides a yellow dye.

15.2 RED MULBERRY (*M. rubra* L.)

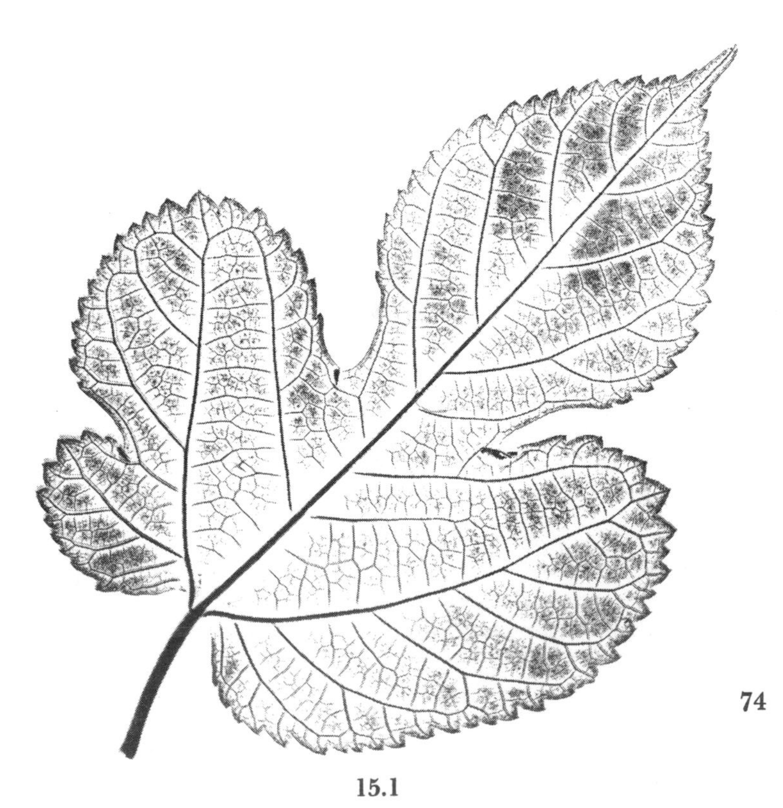

15.1

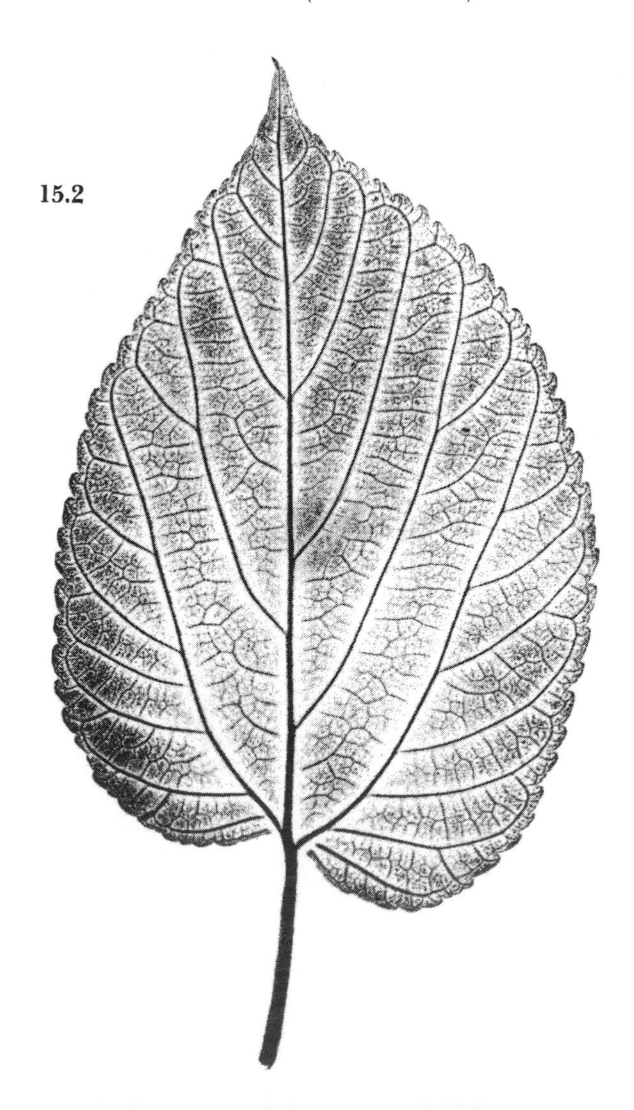

15.2

15.3

THE PAPER-MULBERRY GENUS
(*Broussonetia* L'Her.)

About 2 species in the genus. Fruits are a collection of orange-red drupelets.

15.3 PAPER-MULBERRY (*Broussonetia papyrifera* L'Her.)

This is a tree with a history. In the South Sea Islands tapa cloth is made from it. This material is fashioned into clothing by the natives, but Europeans favor it for backing material for paintings. Nicholas Longworth introduced the tree to Cincinnati as food for silkworms and it has run wild. Only male trees were to be found in Cincinnati, but the odd, insipid fruits are available on trees in Miami. The bark is more or less pink in color and has interesting flow lines, suiting it to use for wall plaques.

15.3

THE OSAGE-ORANGE GENUS
(*Maclura* Nutt.)

Only one species occurs in this genus and its fruit, a syncarp with embedded drupelets, is as large as an orange. The tree has strong thorns.

15.4 OSAGE-ORANGE (*Maclura pomifera* Schneid.)

Its wood is one of the toughest, heaviest, hardest, and most durable of the native hardwoods. It is used for making archery bows. Also its bright orange wood yields a dye of orange-yellow and golden color that was known to the Indians and is still used. Fine white flax-like fibers can be obtained from its inner bark.

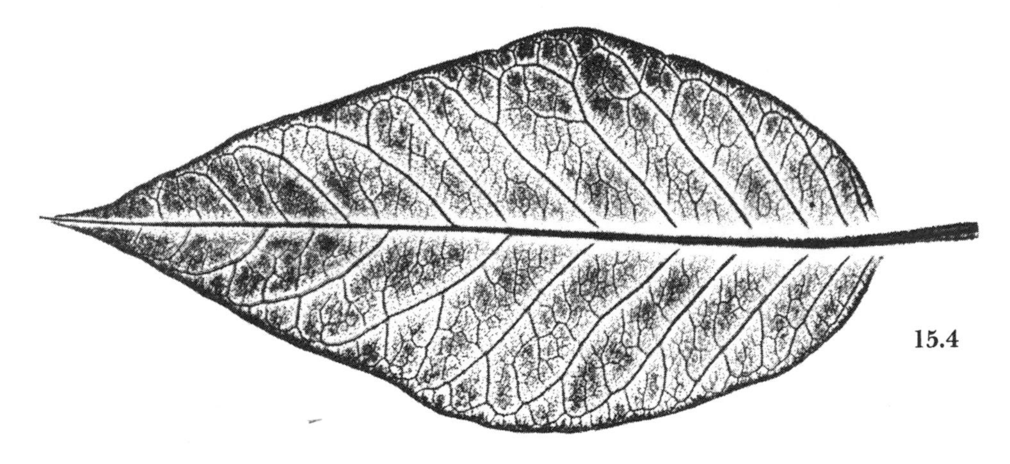

15.4

THE FIG GENUS (*Ficus* L.)

The genus is comprised of nearly 2,000 species of trees, shrubs, and climbers (some of which strangle their supporting host). The fruit is botanically called a synconium, that is, a roundish, expanded receptacle, bearing on its inside minute flowers whose ovaries, after pollination and fertilization, develop into small achenes.

15.5 COMMERCIAL FIG (*Ficus carica* L.)
As the leaf-print indicates, there is some similarity of this to the MULBERRY leaf. There are about a score of other FIGS planted as ornamentals in the southern states or in tubs wintered indoors in the North. This FIG, when planted in colder climates, is often buried for the winter. The dried fruits are an important market commodity and the fresh fruits are sometimes available. The white latex is poisonous. According to Lyons, the dried fruit has been roasted as a substitute for coffee.

15.6 SACRED FIG; BO-TREE (*Ficus religiosa* L.)
A sacred tree of India.

15.7 WALL-FIG (*Ficus repens* of horticulture)

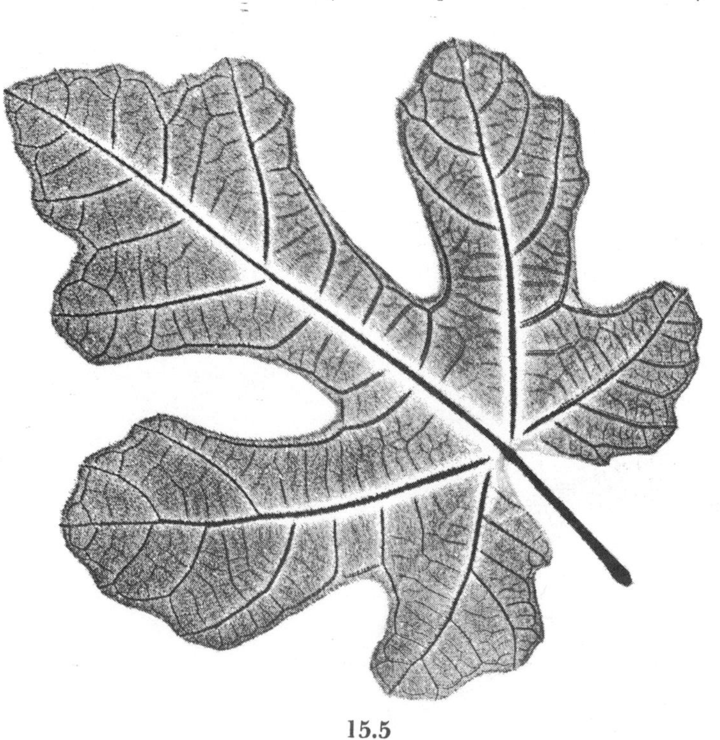

15.5

15.6

15.7

A family of 50 or more genera and about 1,000 species. Many are ornamental plants and one, MACADAMIA, provides the edible nut of commerce.

Many species, some with large showy flowers, are native to the Cape Province of the Republic of South Africa.

THE GREVILLEA GENUS
(*Grevillea* R. Br.)

The 200 or so species of the genus are mostly native in Australia.

16.1 GREVILLEA (*G. Banksii* R. Br.)

The leaves may be 6 to 12 inches long. It may be grown out-of-doors in warm regions, or indoors in tubs further north.

SILK-OAK (*G. robusta* Cunn.)

A valuable timber tree in Australia, but seen here mostly as a small shrub with hoary or rusty-haired branches.

16.1

This family consists of only one monotypic genus of deciduous trees with short spurs on the branches. The leaves are opposite on the shoots but solitary on the spurs. The fruit is a dehiscent pod bearing many winged seeds.

17.1 KATSURA-TREE (*Cercidiphyllum japonicum* Sieb. & Zucc.)

This bushy tree is one of the few that have three distinct foliage color changes during the season. The leaves are purplish when young, become green, and then in the autumn turn bright yellow, sometimes with a touch of scarlet. This is another example of a tree with winged seeds whose yellow leaves may be signaling rodents to the feast.

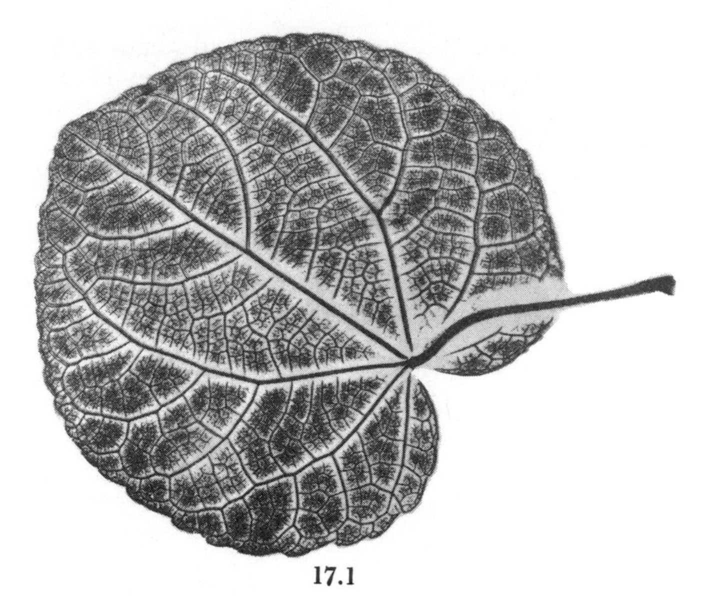

17.1

There are 8 genera and about 20 species of mostly climbing, woody, shrubs in this family. A few of them are valued ornamentally in the U.S.

THE AKEBIA GENUS (*Akebia* Decne.)

The two species of twining shrubs from China and Japan are grown as porch and arbor vines, since they afford good shade. Hand cross-pollination of the flowers is required for good production of their large oblong berries.

18.1 THREE-LEAVED AKEBIA (*Akebia trifoliata* Koidz.)

18.2 FIVE-LEAVED AKEBIA (*A. quinata* Decne.)

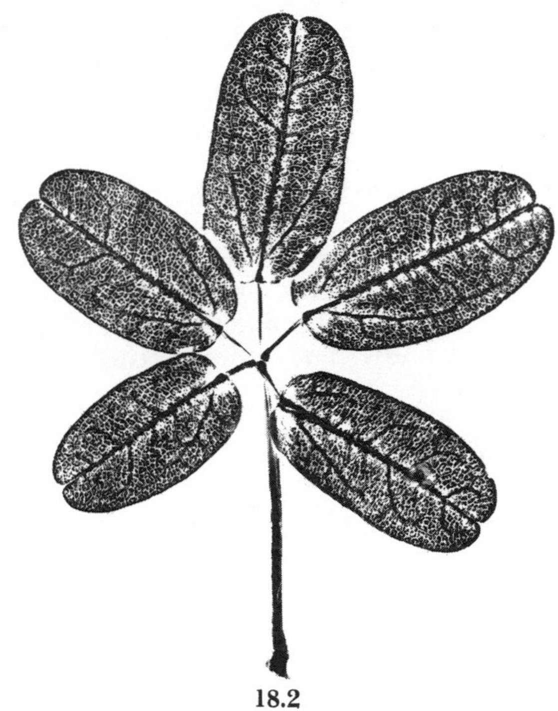

18.2

18.1

About 200 species grouped in 10 or more genera are included in this family of herbs and shrubs. The fruit is a berry or a capsule. To all appearances this heterogeneous family is actually two: The BARBERRY group of woody plants and the MAY-APPLE group of herbaceous plants. Some taxonomists actually separate them that way.

In the BARBERRY group the stems are woody, and this wood is usually yellow in color. The MAHONIAS are here included with the BARBERRIES. They have compound leaves and, in the species that have come east, blue berries. In the JAPANESE BARBERRY the flowers have a mechanically tripped pollination device. I am not certain how far this contrivance extends among its relatives.

THE BARBERRY GENUS (*Berberis* L.)

The leaves are simple and the branches spiny (the spines being modified leaves) . The genus consists of nearly 175 species.

19.1 WINTERGREEN BARBERRY (*Berberis Julianae* Schneid.)

The fruits are black and usually contain one seed. The flowers occur in fascicles of about 15. Introduced from China.

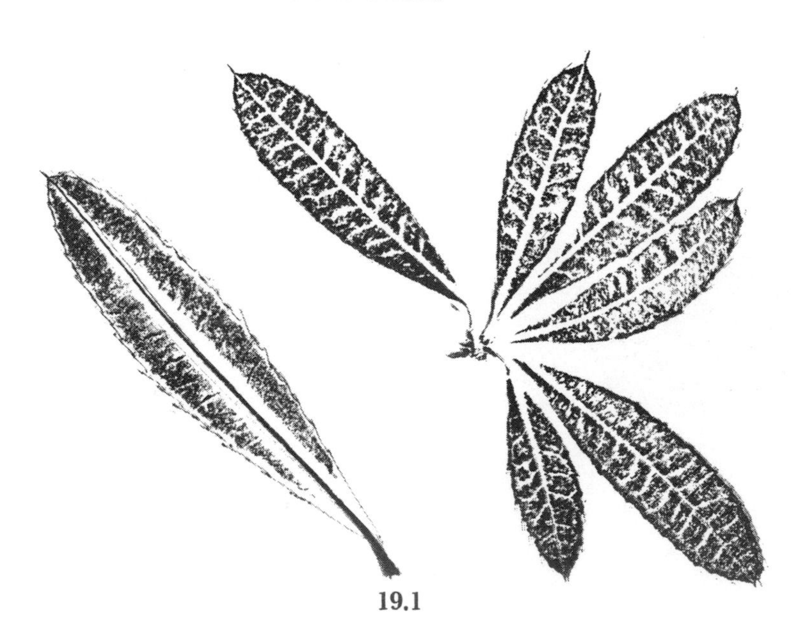

19.1

19.2 ENGLISH or "COMMON" BARBERRY (*Berberis vulgaris* L.)

Once called COMMON BARBERRY, this shrub is very uncommon in America today. It is the alternate host of the wheat rust fungus, and was almost exterminated by the USDA for this reason many years ago. The red berries of this species were made into jams and jellies by the colonists. In Europe an infusion of the roots was believed to be good for chronic dyspepsia. The leaves have a high content of vitamin C.

19.2

19.3 JAPANESE BARBERRY (*B. Thunbergii* DC)

This species is now the more common of the two. It does *not* harbor the wheat rust fungus. It is widely used as a decorative hedge and occurs in both green and reddish-leaved forms.

19.3

THE MAHONIA GENUS (*Mahonia* Nutt.)

There are about 50 species of evergreen shrubs noted for their beautiful foliage and yellow flowers included in this genus. They differ from BARBERRY in their lack of spines and in having pinnately compound leaves.

19.4 HOLLY MAHONIA (*Mahonia Aquifolium* Nutt.)

An evergreen shrub of the Pacific Northwest that has become popular for ornamental planting in the eastern states. The yellow flowers are a minor source of honey. The bluish berries are edible raw or cooked and could perhaps be improved by upbreeding for commercial purposes.

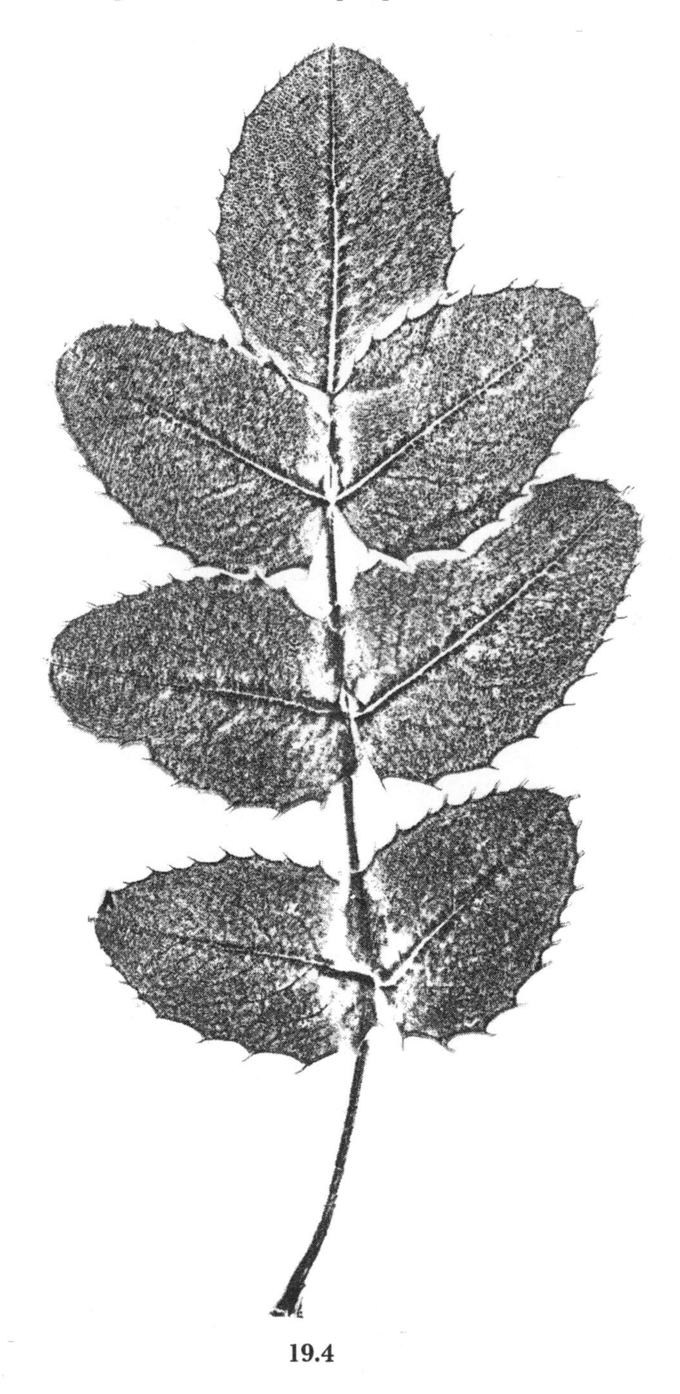

19.4

The tropics, subtropics, and temperate zones of North America and Asia are home to the 10 genera and 80 or more species of this family. Many of these deciduous trees and shrubs are highly ornamental.

There are many MAGNOLIAS, but only one TULIP-TREE (except that our tree has a cognate in the Orient and ancestors in the rocks). As in the WILLOWS and POPLARS, we can in this family see the effect of the environment on evolution. Both the TULIP-TREE and the MAGNOLIAS have more or less showy flowers pollinated by insects. The MAGNOLIAS, of smaller stature (putting them in the ornithosphere, the area closer to the earth and inhabited by birds), have red berries. The much taller TULIP-TREE, exposed where the "winds hold sway," has dry-winged seeds or, technically, fruits.

The MAGNOLIA family is considered most primitive among the generally woody plants, just as the BUTTERCUPS are held to be among the generally herbaceous ones. But the MAGNOLIAS are advanced to the extent that their fruit pods are compounded and have a spiral rather than radiate arrangement. The berries of the true MAGNOLIAS hang from slender threads, a convenience for the birds.

Some of these trees are of the highest ornamental merit. The EVERGREEN MAGNOLIA may well be the most beautiful of trees. The TULIP-TREE, which was merely mentioned in passing in the 1924 edition of Bailey's *Manual of Cultivated Plants,* has come to be the favorite of many people.

THE MAGNOLIA GENUS (*Magnolia* L.)

The rather large flowers bear 6 to 12 petals and may be white, yellow, rose, or purple. In the 35 species, the leaves, which may be deciduous or evergreen, are unlobed and aromatic. The berries are red.

20.1 BIGLEAF or DINNERPLATE MAGNOLIA (*Magnolia macrophylla* Michx.)
Tip of an enormous leaf.

20.2 EVERGREEN MAGNOLIA (*M. grandiflora* L.)
Thick, glossy leaves with brown scurf beneath.

20.3 UMBRELLA MAGNOLIA (*M. tripetala* L.)
Large leaves, clustered at tip of twig.

20.4 SPRING, SAUCER, or SOULANGE MAGNOLIA (*Magnolia Soulangeana* Soul.)
The Magnolia of the nursery trade. Often a shrubby plant, its bark is gray. It is a hybrid, one of whose parents is the following.

20.5 CHINESE MAGNOLIA (*M. denudata* Desr.)
This is a tree with a single trunk. Note the finer venation of the leaf.

20.1

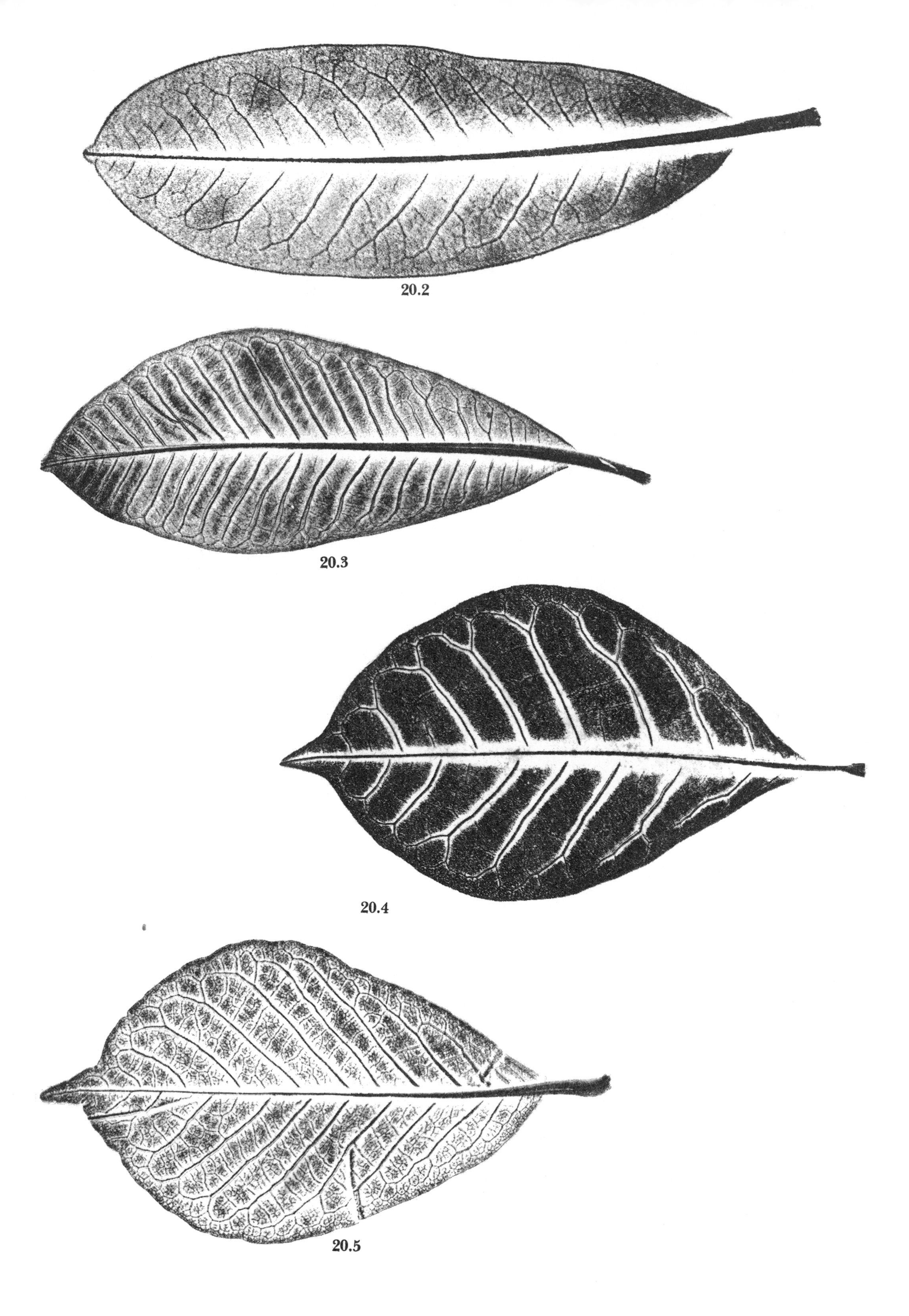
20.2
20.3
20.4
20.5

20.6 STARFLOWER MAGNOLIA
(*M. stellata* Maxim.)

Recognizable because of the rather small wrinkled leaves and the many-petaled flower.

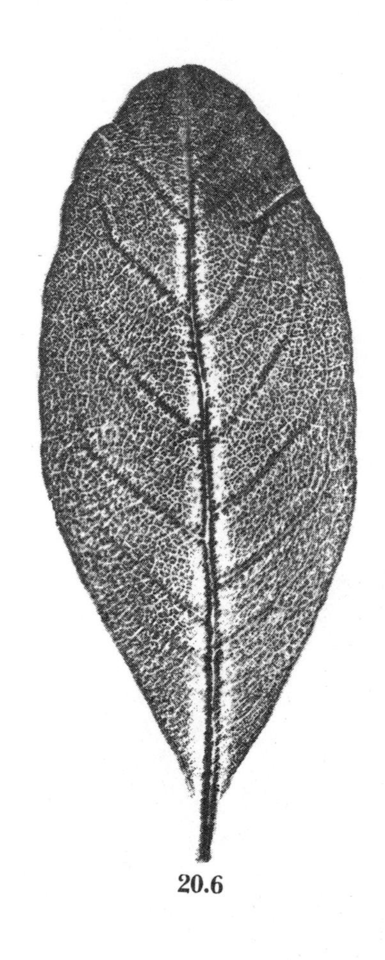

20.6

20.8 CUCUMBER-TREE (*M. acuminata* L.)

The fruit is a 3- to 4-inch pink or red "cucumber." One of the most abundant Magnolias, its light, soft, but durable wood is used for millwork, boxes, and cheap furniture.

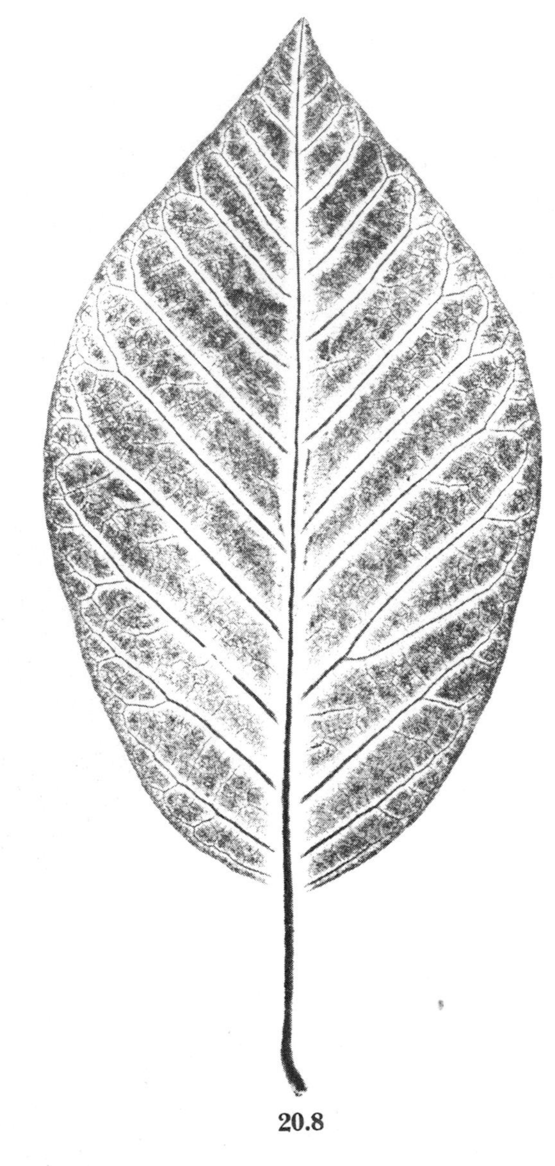

20.8

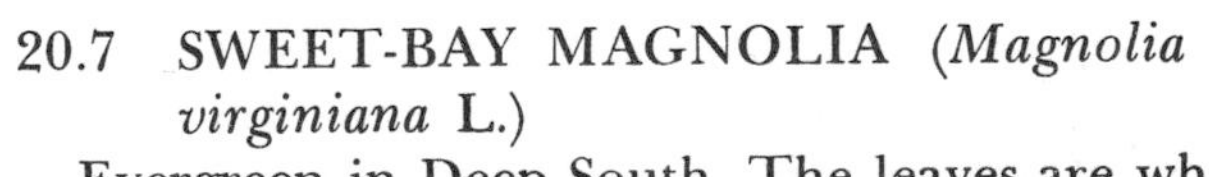

20.7 SWEET-BAY MAGNOLIA (*Magnolia virginiana* L.)

Evergreen in Deep South. The leaves are white-waxy beneath.

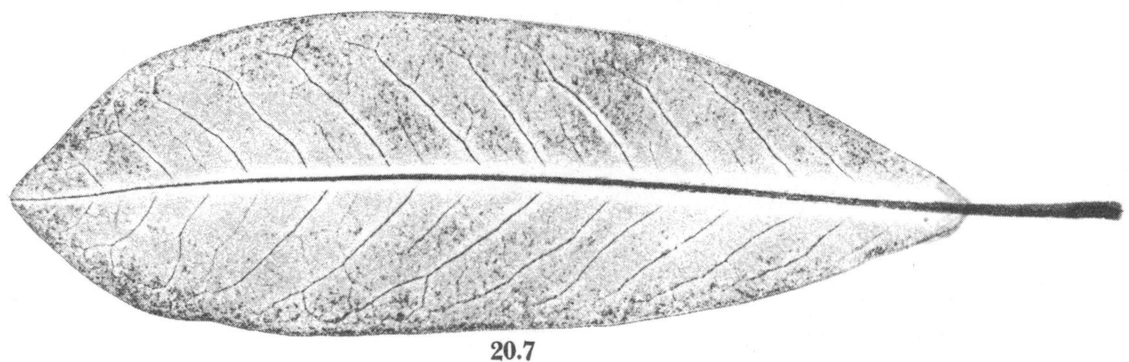

20.7

THE TULIP-TREE GENUS
(*Liriodendron* L.)

One of the largest of our trees, being 125 to 250 feet high and with diameters of 6 to 14 feet. The wood is soft, light, easily worked, with fine straight grain, stiff and durable, but not very strong. It is especially valued for the excellent way in which it accepts glue. The leaves are waxy and water-shedding, turning a clear rich yellow in the fall. There are but two species in the genus.

20.9 TULIP-TREE (*Liriodendron tulipifera* L.)

20.9

20.10 CHAMPAC (*Michelia Champaca*)

Oil of Champaca, one of the most famous perfumes of India, is extracted from the large yellow flowers, which are also worn by the natives as ornaments.

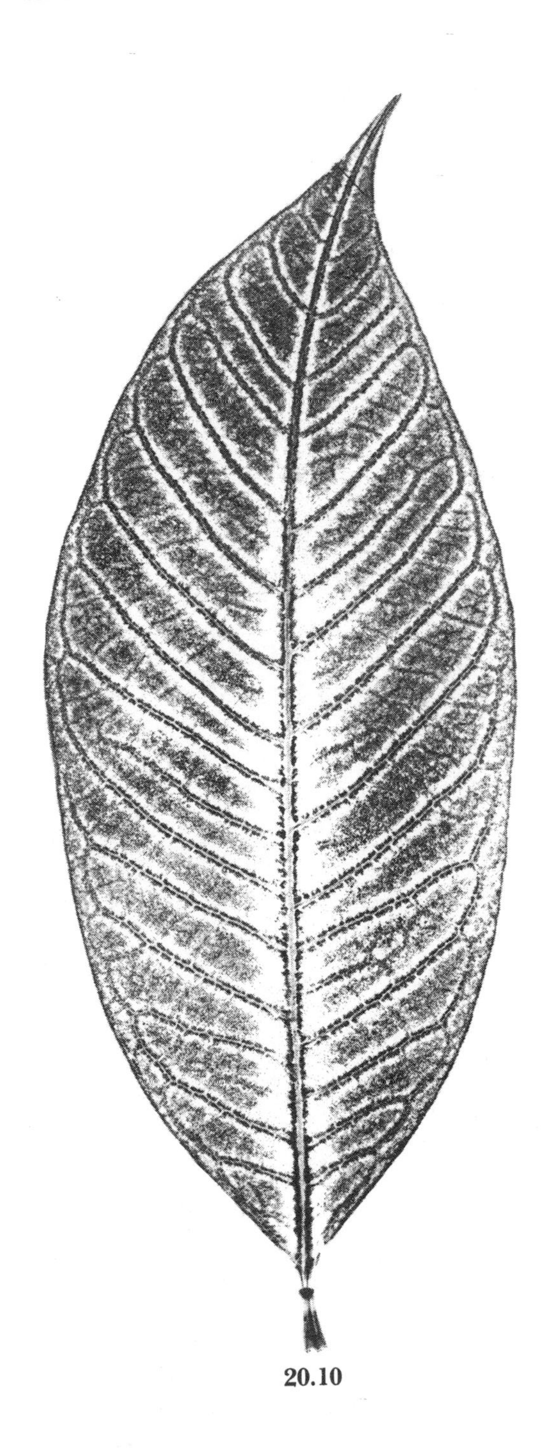

20.10

The approximately 600 species in the 40 to 50 genera of this family are mostly tropical shrubs and trees of the Old and New World, although some extend into temperate regions.

The term "custard-apple" has been applied to the fruits of various species in the genus *Annona*. These fruits are fleshy syncarps (as are pineapples), formed by the fusion of many ripened ovaries and the receptacle of the flower. The CHERIMOYA (*A. Cherimola*) is a delicious dessert fruit with a very aromatic white or yellowish flesh. It is soft and custard-like.

21.1 CHERIMOYA (*Annona Cherimola* Mill.)

In addition to its edible fruit, the tree is noteworthy for the fibers derived from its inner bark, which have been used in the weaving of fish-nets.

21.2 PAPAW-TREE (*Asimina triloba* Dunal)

A small tree with dark purple axillary flowers giving rise to edible fruits that are 3 to 7 inches long and 1 to 2 inches thick. They are yellow at first and later turn brown. The tree grows from Ontario and Michigan to Florida and Texas.

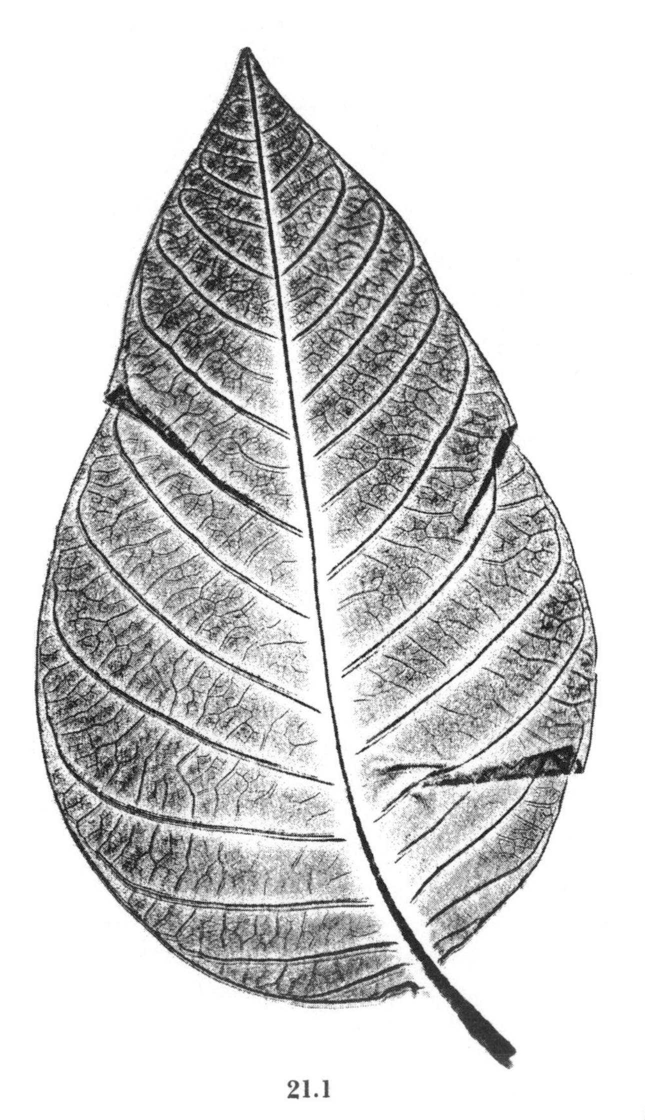

21.1

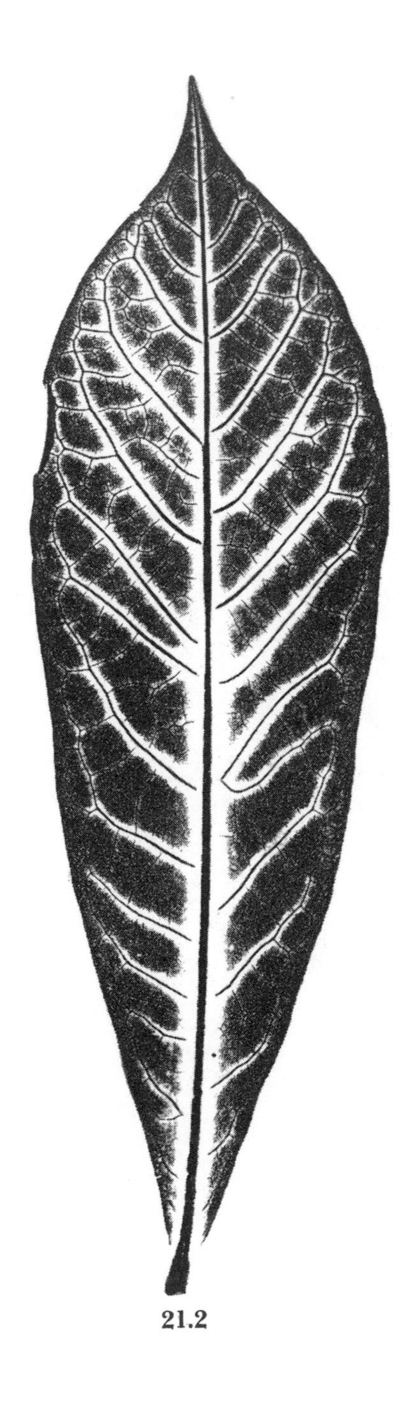

21.2

AVOCADO, CINNAMON, and CAMPHOR are well-known articles of commerce that are derived from three among the 1,000 species (in about 40 genera) of aromatic trees and shrubs in the LAUREL family. In most cases the leaves of these trees are leathery and evergreen, although some are soft, thin, and deciduous. The fruits are indehiscent berries or drupes.

THE AVOCADO GENUS (*Persea* Gaertn.)

About 50 species of trees and shrubs with leathery, pinnately veined leaves.

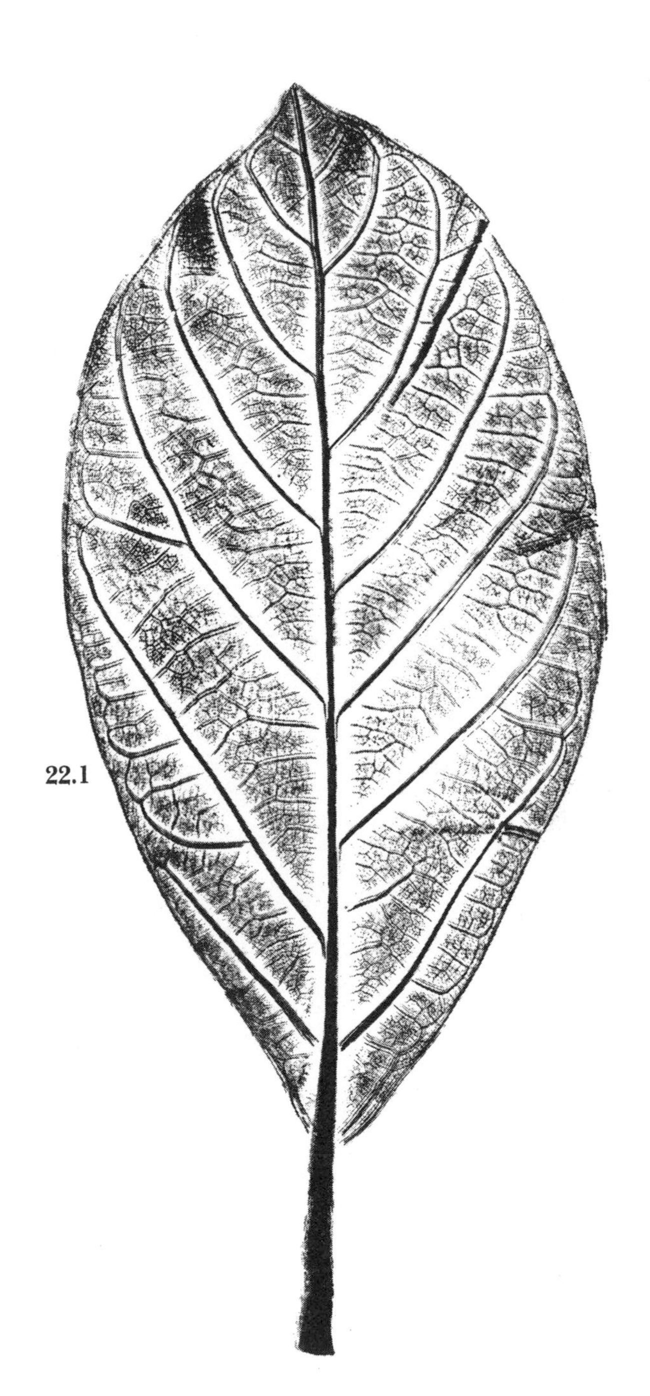
22.1

22.1 AVOCADO (*Persea americana* Mill.)

Formerly the "Alligator-pear," now more commonly called the "Avocado." This fruit is unusual because it contains more protein than any other and a large amount of fat. It is particularly valuable in tropical diets because it is also rich in the vitamin B complex and vitamin A. The pulp is eaten raw, cubed into salads, or used as a sandwich spread.

THE CAMPHOR GENUS
(*Cinnamomum* Blume)

More than 50 species of evergreen trees and shrubs native in Asia and Australia. Noted for their aromatic and medicinal properties.

THE SASSAFRAS GENUS (*Sassafras* Nees.)

Three species of aromatic deciduous trees whose bark is supposed to have medicinal properties and has been used since Colonial days as a spring tonic "to thin the blood."

22.3 SASSAFRAS (*Sassafras albidum* Nees.)
An aromatic tree, famous in camping circles as the source of sassafras tea, made by boiling cut-up pieces of the bark of the roots in water. The young leaves are also edible. The extracted flavoring has been used in tobacco, root beer, soaps, and perfumes, as well as floor and polishing oils.

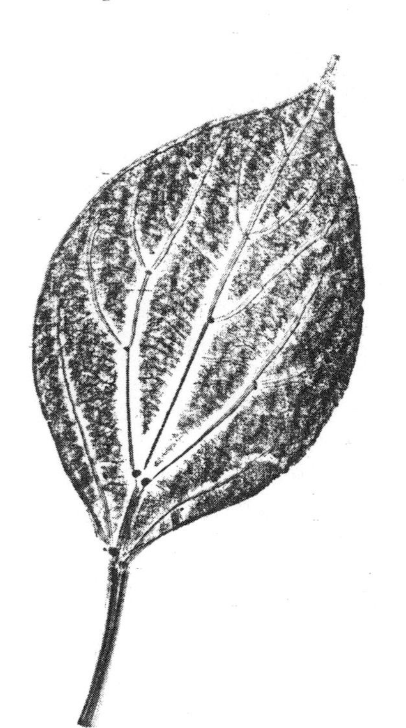
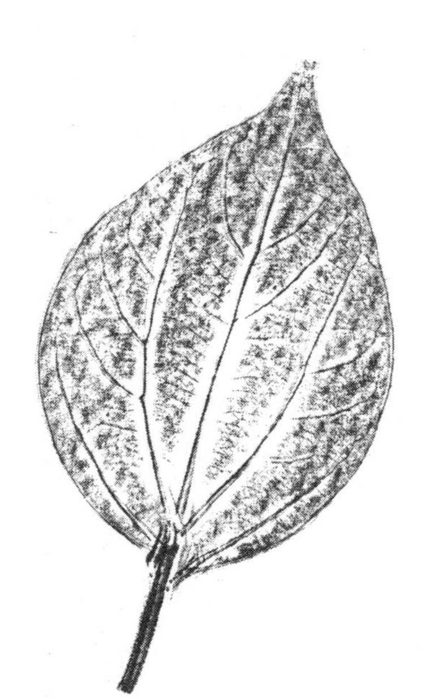

22.2

22.2 CAMPHOR TREE (*Cinnamomum Camphora* Nees & Eberm.)
This tree provides one of the most important essential oils used in industry. *C. zeylanicum* is the source of the spice cinnamon.

22.3

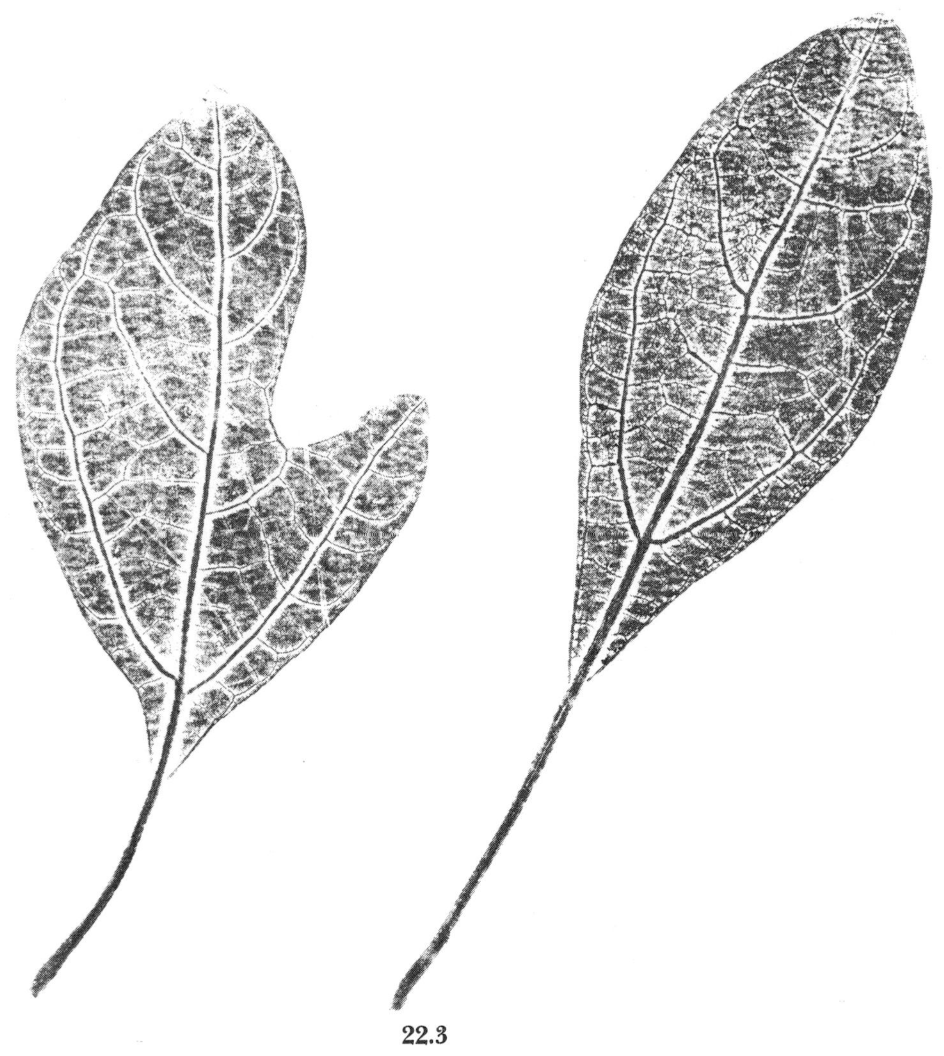

22.3

THE LAUREL GENUS (*Laurus* L.)

Two species only. The fruits are small berries.

22.4 LAUREL; SWEET BAY (*Laurus nobilis* L.)

The famous laurel of history—an evergreen tree with smooth, stiff, dull-green leaves.

22.4

THE SPICE-BUSH GENUS
(*Lindera* Thunb.)

About 60 species of aromatic deciduous or evergreen trees and shrubs.

22.5 SPICE-BUSH (*Lindera Benzoin* Blume)

This is a tall, attractive bush, highly ornamental and especially well adapted for use as a natural hedge in dense shade. The spring-appearing yellow flowers are fragrant. The red berries and their seeds are used as spices. The twigs are boiled to make a palatable purple-colored tea. According to Johnson, the bark and berries have diaphoretic and vascular stimulant properties. It has also been used in Appalachia for dysentery, coughs, and colds. Note that while in the related SASSAFRAS the berries are blue and the leaves turn red, in SPICE-BUSH the berries turn red and the leaves yellow.

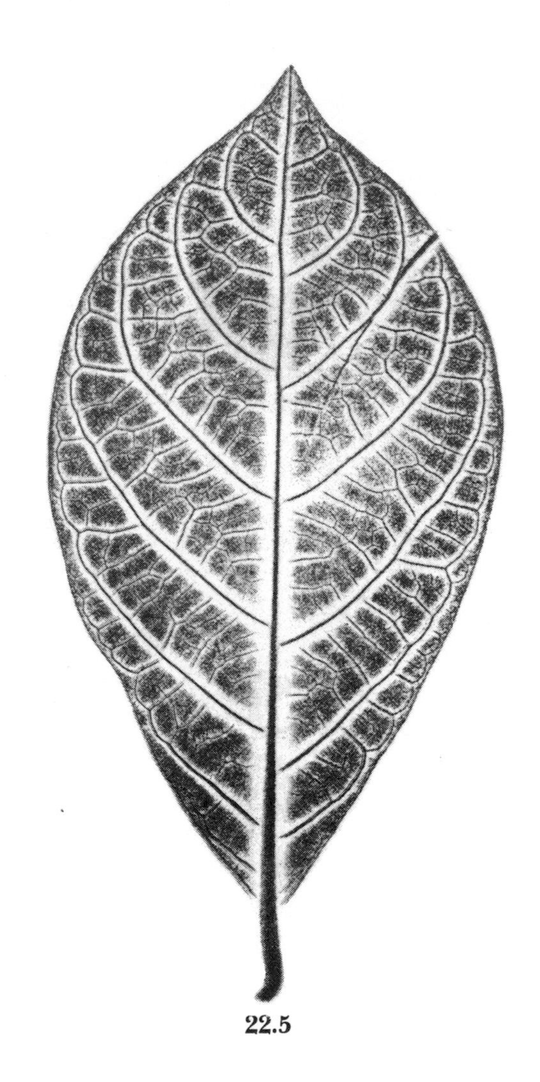

22.5

A very mixed family that includes herbs, shrubs, and pomological species that inhabit both temperate and subarctic regions. There are about 70 to 100 genera and between 900 to 1,100 species, depending on the taxonomic viewpoint. These plants are allied to the Rose family. The fruit is a capsule or berry having many seeds and abundant albumen. CURRANTS, GOOSEBERRIES, HYDRANGEA, and MOCK-ORANGE are rather well-known members of the family.

THE MOCK-ORANGE GENUS
(Philadelphus L.)

About 30 species of wide distribution which, through horticultural manipulation, have become taxonomically confused.

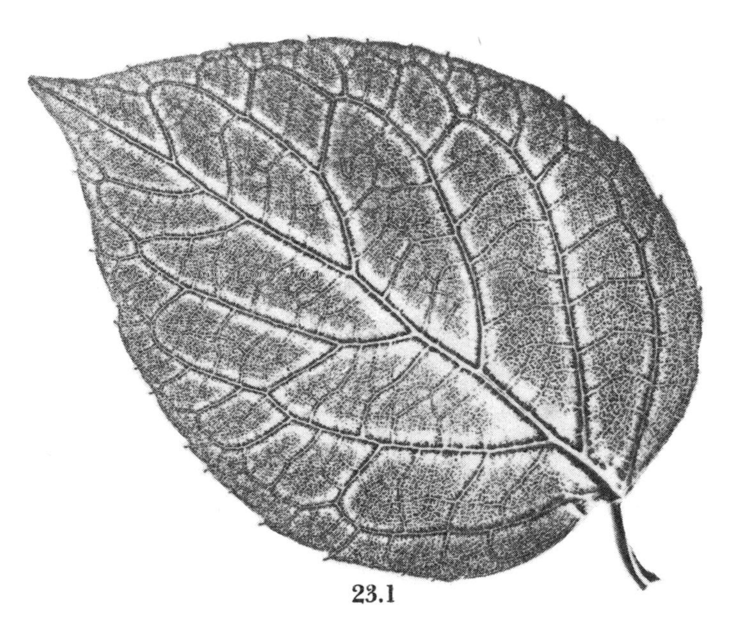

23.1

23.1 COMMON MOCK-ORANGE; "SYRINGA" *(Philadelphus coronarius L.)*
Popular in cultivation because of the snowy white and very fragrant blooms. The genus name is after King Ptolemy Philadelphus in the third century B.C. The name "Syringa" should not be used for this species.

Some 100 species of trees and shrubs have been assigned to about 23 genera comprising the family.

THE WITCH-HAZEL GENUS
(*Hamamelis* L.)

About 6 species are found in eastern North America, China, and Japan.

24.1 WITCH-HAZEL (*Hamamelis virginiana* L.)

Usually planted for their yellow flowers, which bloom during the winter, that is, from autumn to early spring. In the wild this species is often found on lake shores and river banks. Straight sections of the stem that are about 1½ to 2 inches in diameter and perhaps 2 feet long are ideal for the handcrafting of fireplace brooms. The wood and a penknife are all that are needed. See Mason for instructions. The forked branches are a favorite source of inspiration for "water-dousers."

THE SWEET-GUM GENUS
(*Liquidambar* L.)

Four species of deciduous trees in North and Central America and Asia.

24.2 SWEET-GUM TREE (*Liquidambar Styraciflua* L.)

A very beautiful tree whose leaves turn deep crimson in the fall. The fruit is a long-stemmed, round head, useful in decorative craftwork. SWEET-GUM wood is light, soft, tough, and resilient, with reddish-brown, fine, straight, close grain. It polishes well. Bruising the bark of this tree causes the production of a balsam in the inner bark, which is recovered by boiling in sea water. The resulting aromatic grayish-brown substance is called "Styrax." It has been used in soaps and cosmetics, and in medicine to stimulate the mucous membranes and to treat scabies.

24.1

24.2

EUCOMMIACEAE

An interesting family of only one genus. The 60-foot or higher deciduous tree is similar to Elm in appearance, but it produces rubber latex beneath the bark of twigs and branches. It is a native of China.

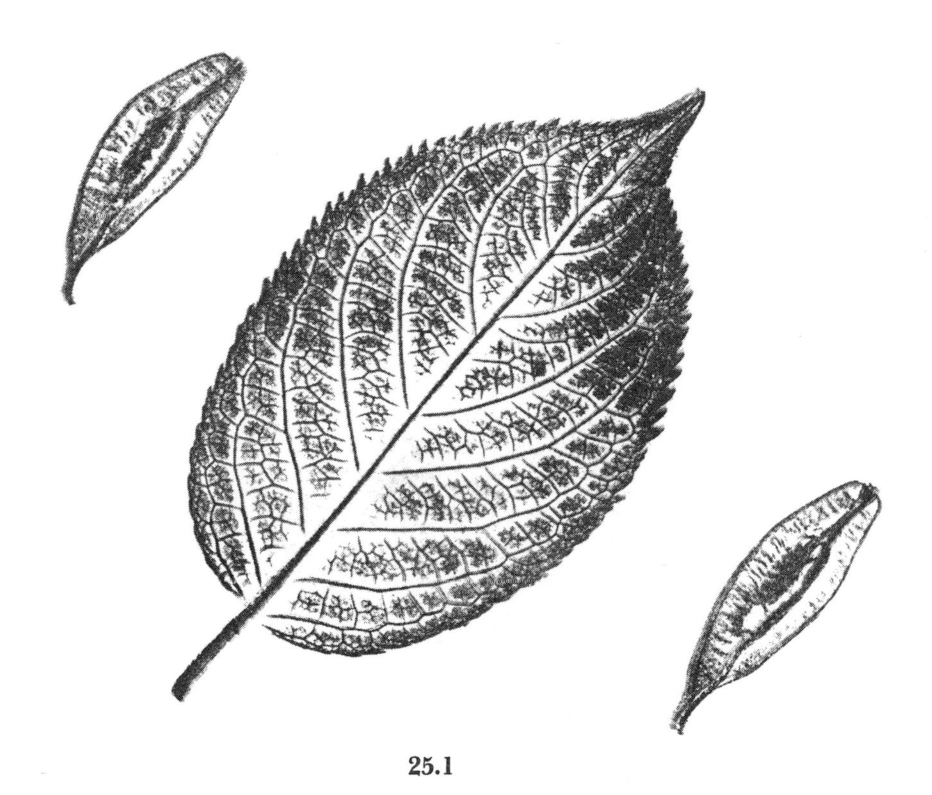

25.1

25.1 HARDY-RUBBERTREE; TRUEGUM-TREE (*Eucommia ulmoides* Oliv.)
The hanging fruits have dark-brown wings.

There are about 6 species of large monoecious trees in the single genus that makes up the *Plane-tree* family. The bark in the two species shown below is generally almost white, but with a greenish or yellowish cast. The branches may be white and more or less mottled with gray patches. The bark of the upper trunk and the branches peels away in large plates. Although the name is widely used, these trees should not be called "Sycamores," since that name already belongs to one of the FIGS (*Ficus Sycamorus*).

26.1 BUTTONWOOD; AMERICAN PLANE (*Platanus occidentalis* L.)

The seed-balls are single. This is one of the largest hardwood trees in the U.S., with wood that is hard, tough, strong, and very durable. It has a close, uneven grain which, when quarter-sawed, may be marketed as "lacewood."

26.2 LONDON PLANE (*Platanus acerifolia* Willd.)

The seed-balls are usually in twos. This tree is widely planted as a shade tree since it is tolerant of city smoke and alkali.

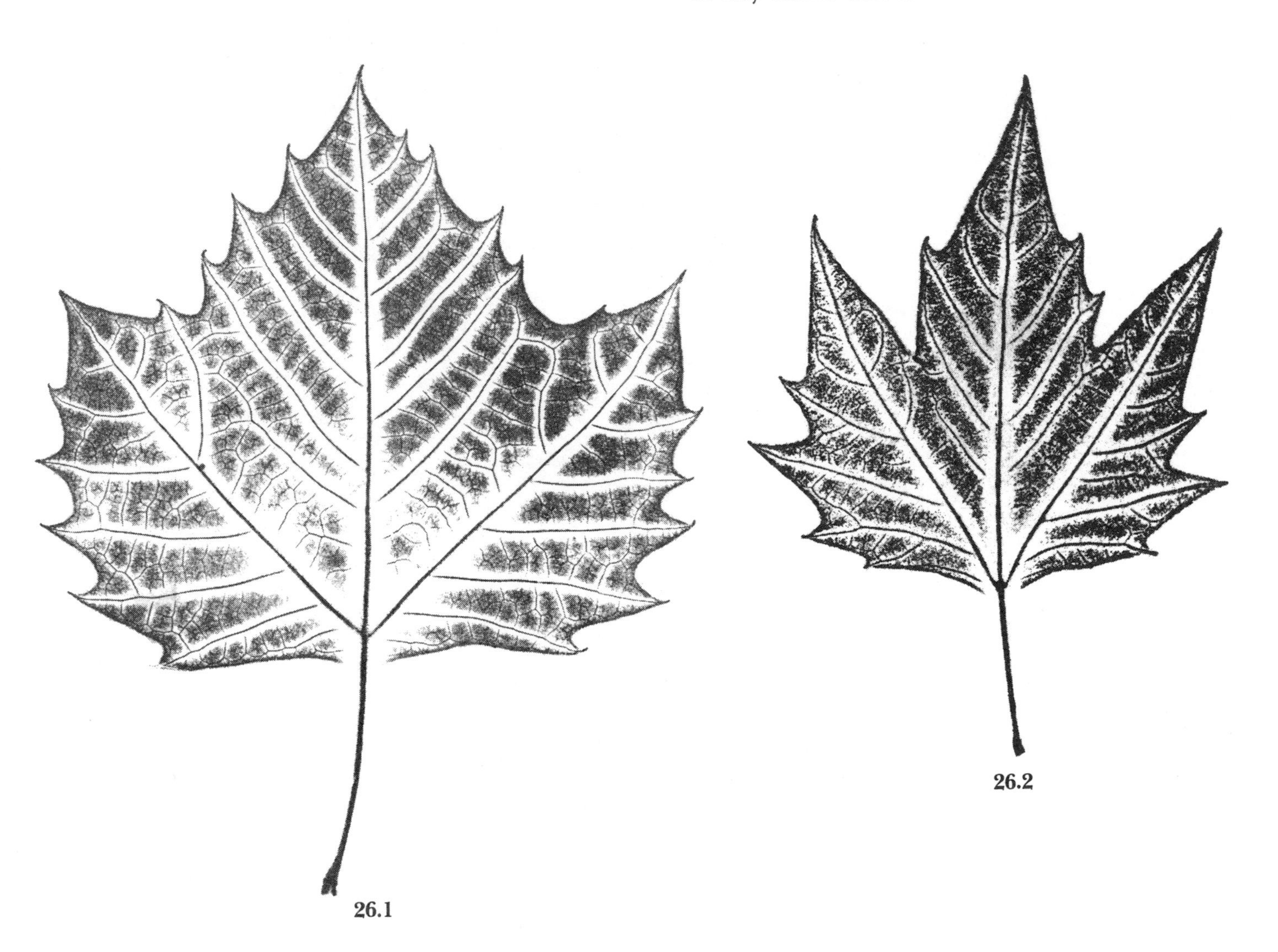

26.1

26.2

This family of herbs, shrubs, and small trees, some spiny and some not, consists of about 115 genera and 3,200 species. The fruits are quite varied in the family since achenes, follicles, hips, pomes, and drupes all occur. Doubtless this is the most important pomological family in the world, completely dominating the temperate zones. It would be hard to picture a world without apples, pears, strawberries, peaches, and the like.

The flowers are of a type normally pollinated by bees and they yield enormous quantities of high-quality honey. Looked at another way, they carry the bees through the spring and perpetuate their pollinating propensities for the rest of the year to the benefit of countless other flowering plants.

Not all of the fruits in this family are edible. Some are dry pods and a few are burs. But those that are edible have long been, and will doubtless continue to be, of very special interest to hybridizers.

Many members of this family contain a high percentage of tannin and are therefore astringent (remember, not to be confused with bitter). Tannin is a very important chemical, the chief preservative of leather and it has been an essential medicine against the suppuration that often follows burns. It also has other uses in both medicine and industry.

One should not fail to mention the ROSES and many other members of this family, which are of surpassing beauty.

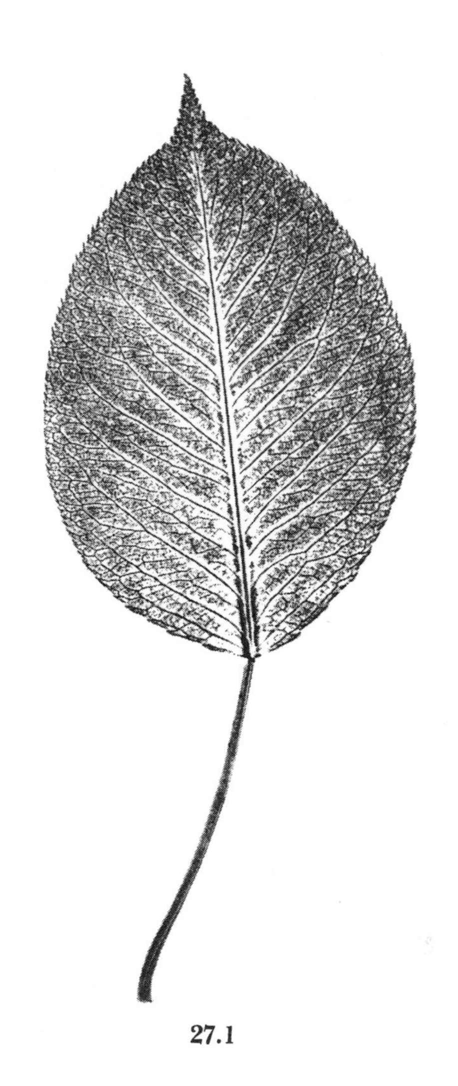

27.1

THE PEAR GENUS (*Pyrus* L.)

Possibly 20 species of deciduous or half-evergreen small trees grown in orchards. They are North African and Eurasian in origin. The pears differ from the apples in that their leaves are more glabrous (hard and glossy) at maturity. Also, the fruit of the pear contains many stone-cells.

27.1 PEAR-TREE (*Pyrus communis* L.)
Recognizable in the distance by its pyramidal form.

THE APPLE GENUS (*Malus* Mill.)

This genus, which is by some considered a subgenus of *Pyrus,* consists of about 25 species. The leaves are softer than in the pear; they have acutely serrate margins, and there are no stone-cells in the flesh of the fruit. Stone-cells (sclereids) are clusters of very heavy-walled cells that give the flesh of the pear its characteristic grittiness. They are not present in the apple.

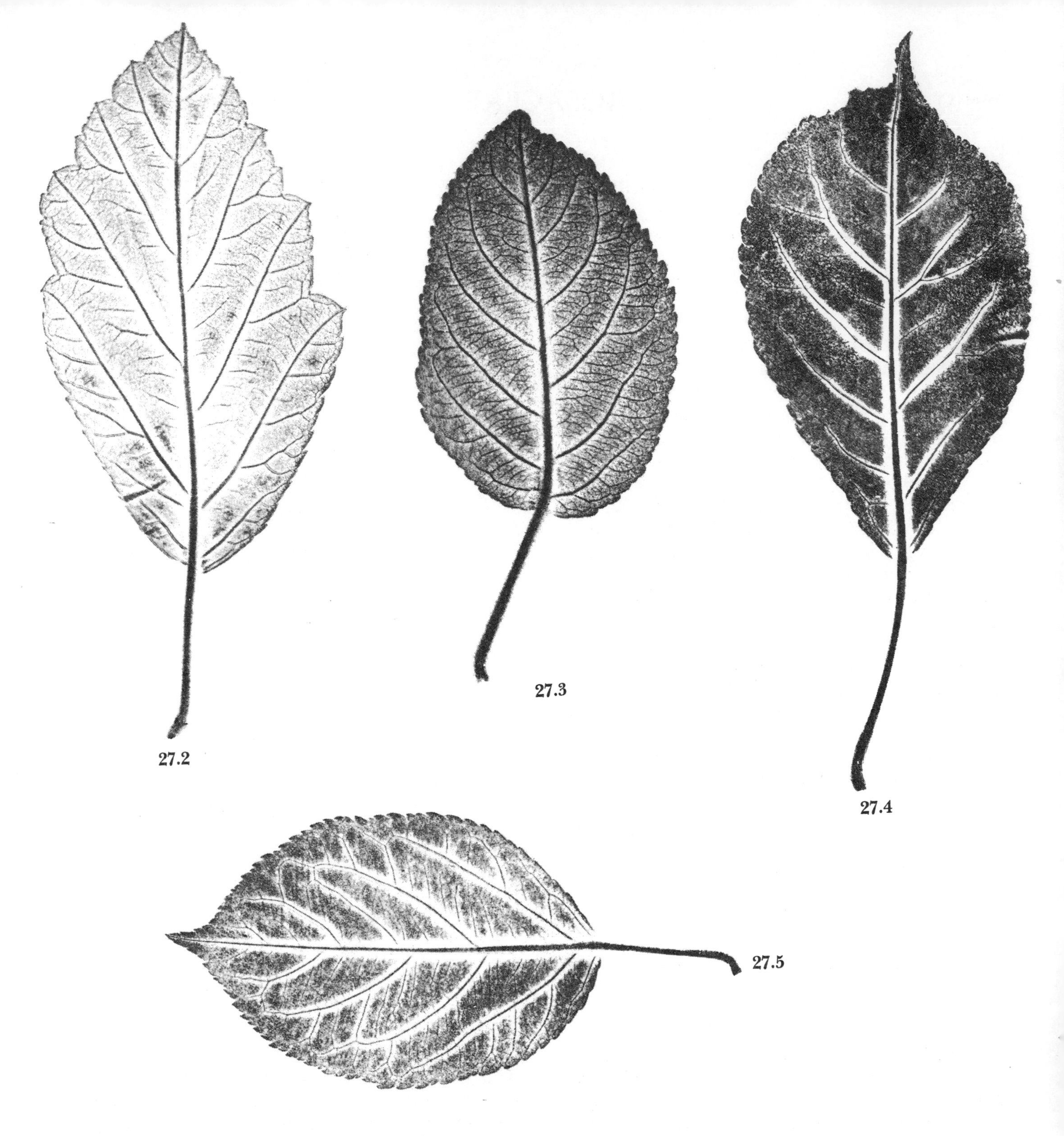

27.2 BECHTEL'S CRAB or ROSE TREE (*Malus ioensis* Britt.)

The cultivated form has double flowers.

27.3 COMMON or ORCHARD APPLE (*M. pumila* Mill.)

Throwbacks with small fruits may be called "Crabs."

27.4 PURPLE CRAB (*M. pumila* var. *Niedzwetzkyana*)

The fruit and underveins of leaves are tinged with purple.

27.5 FLORIBUNDA CRAB (*M. floribunda* Sieb.)

THE STONE-FRUIT GENUS (*Prunus* L.)

A major group of pome-bearing and ornamental plants consisting of approximately 150 to 175 species. Plums, cherries, almonds, apricots, and peaches are included. The edible part of the fruit, in most instances, is the succulent outer flesh, an exception being the almond, where the flesh is dry and the seed alone is eaten. Not all members of the genus have edible fruits; some are too bitter.

The twigs, wilted leaves, and the kernels of the seeds of this genus should be considered dangerous because of the hydrocyanic acid that may be present.

27.6 PEACH-TREE (*Prunus Persica* Batsch)
Young trees, when seen in a nursery, can be recognized by their green-and-red twigs.

27.7 WILD-GOOSE or MUNSON WILD PLUM (*Prunus Munsoniana* Wight & Hedr.)
Distinguished by its troughed leaves. This species is the parent of several horticultural varieties.

27.8 SOUR CHERRY (*Prunus Cerasus* L.)
The cherries are always red. Note glands on petiole.

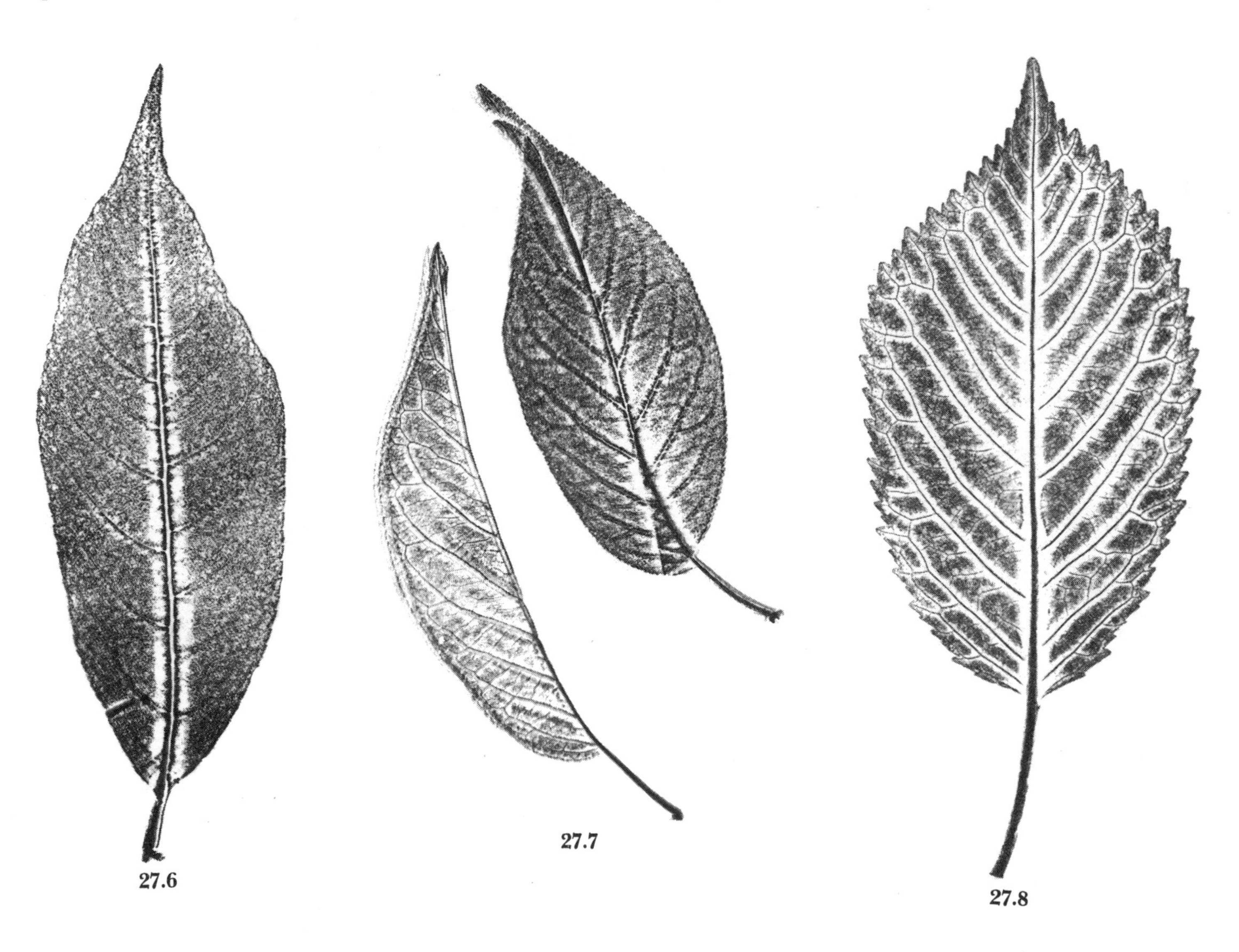

27.6 27.7 27.8

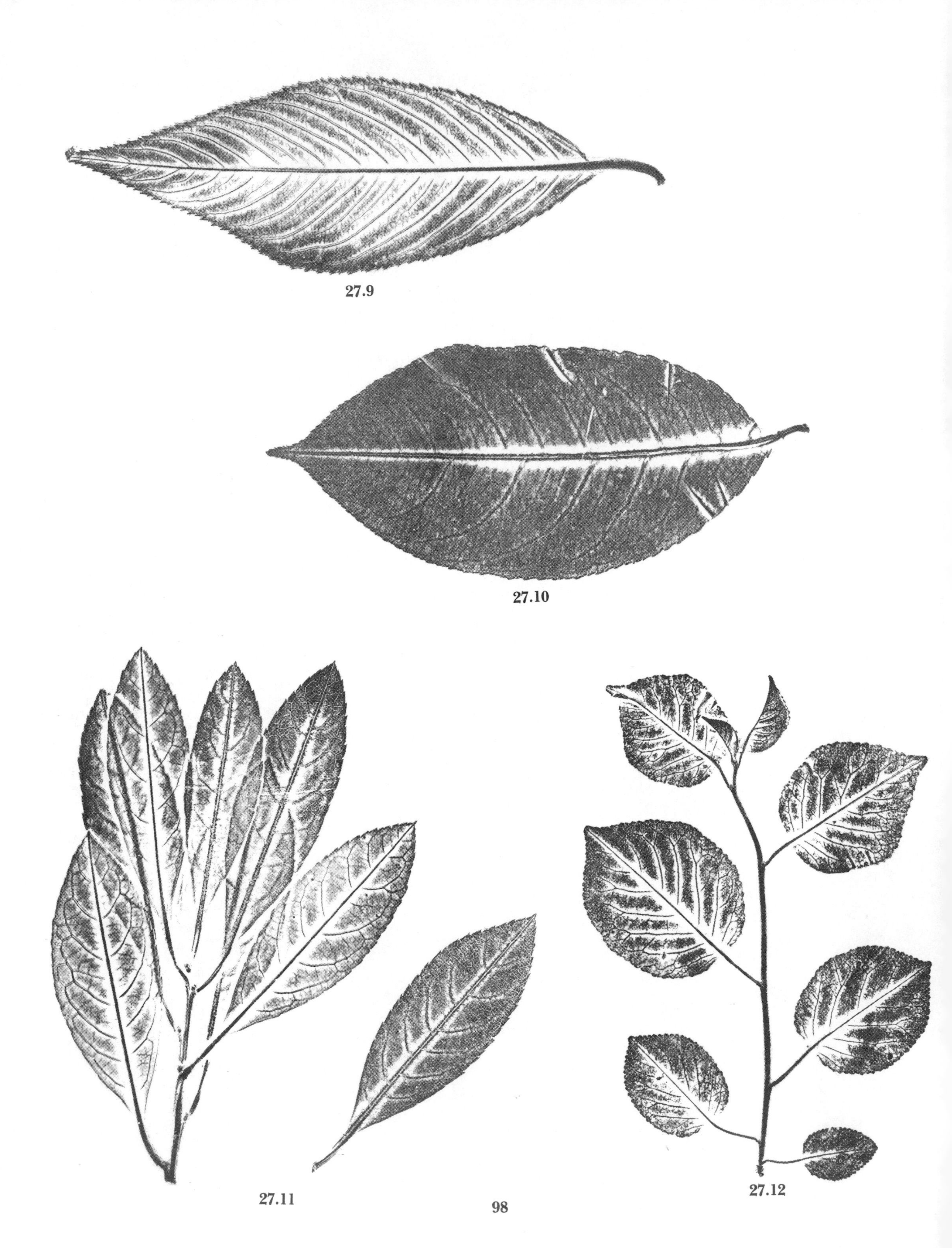

27.9

27.10

27.11

27.12

27.9 WEEPING CHERRY (*Prunus subhirtella* var. *pendula* Tanaka)
Easily recognized by the weeping twigs.

27.10 WILD BLACK CHERRY (*Prunus serotina* Ehrh.)
Chiefly wild. Its flowers and fruits are in racemes. The leaves turn red.

Some 46 species of birds have been recorded feeding on cherry trees.

27.11 SAND CHERRY (*Prunus pumila* L.)

27.12 ST. LUCIE or MAHALEB CHERRY (*P. Mahaleb* L.)
These are chiefly wild trees with small, round leaves. The fruits provide a violet dye and the seeds an oil used in perfumery.

THE KERRIA GENUS (*Kerria* DC.)

Planted here as an ornamental, these native Japanese shrubs are all assigned to a single genus. They are bramble-type bushes, but without thorns. Their golden-yellow flowers range from 1 to 2 inches in diameter.

27.13 WHITE KERRIA; GLOBE-FLOWER (*Kerria japonica* DC.)

27.13

THE LOQUAT GENUS (*Eriobotrya* Lindl.)

Approximately 10 species of East Asian evergreen trees and shrubs are assigned to the LOQUAT genus.

27.14 LOQUAT (*Eriobotrya japonica* Lindl.)
The "Japan Plum" has an agreeably acid fruit that ripens in the spring of the year that follows the blooming of the flower.

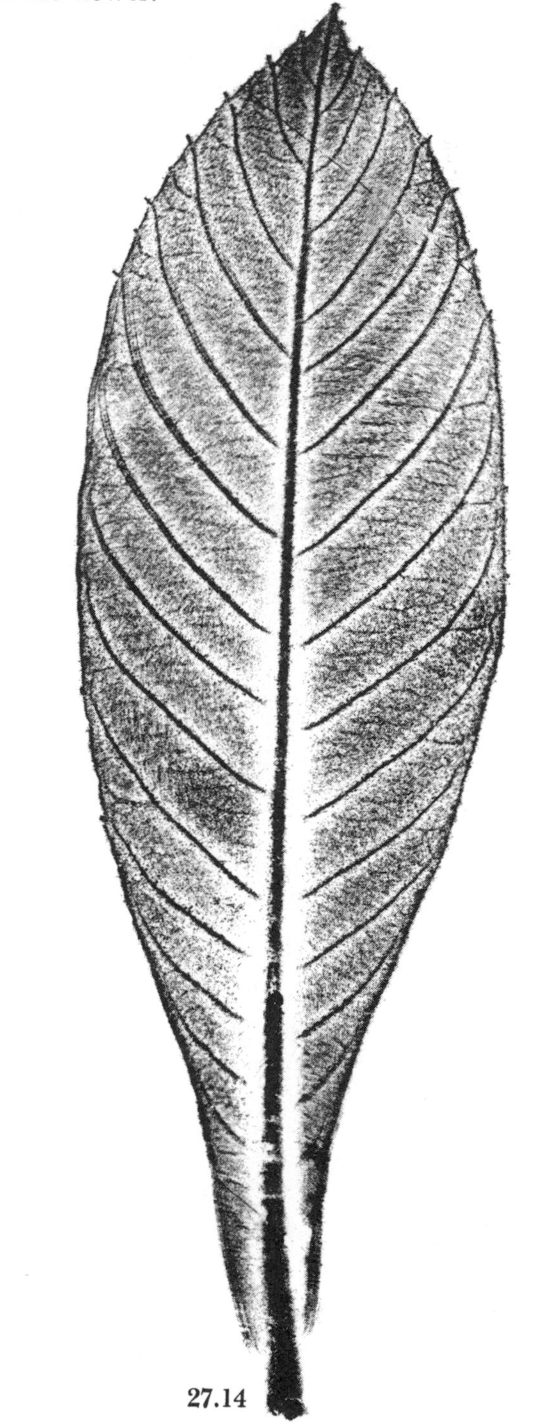

27.14

THE MOUNTAIN-ASH GENUS
(*Sorbus* L.)

In the northern hemisphere there are about 85 species of trees and shrubs belonging to this genus, many of which are grown as ornamentals. The species shown here has bright red fruits that persist into the winter and are attractive to wildlife.

27.15 EUROPEAN MOUNTAIN-ASH; ROWAN-TREE (*Sorbus Aucuparia* L.)

Fifteen species of birds have been recorded feeding on the berries of the ROWAN-TREE. They are also eaten by man in jellies and preserves.

THE FLOWERING QUINCE GENUS
(*Chaenomeles* Lindl.)

The *Flowering Quince*'s three species of usually thorny hardwood shrubs are particularly noted for their beautiful, very early, flowers. The fruits vary among the species from small to large pomes with a closed top (unlike the apple, since the sepals are not persistent). The fruits are more or less enjoyably edible, having been used in times past to make preserves and jellies. They are more tart than the true quinces in the genus *Cydonia*.

27.16 CHINESE QUINCE (*Chaenomeles sinensis* Koehne)

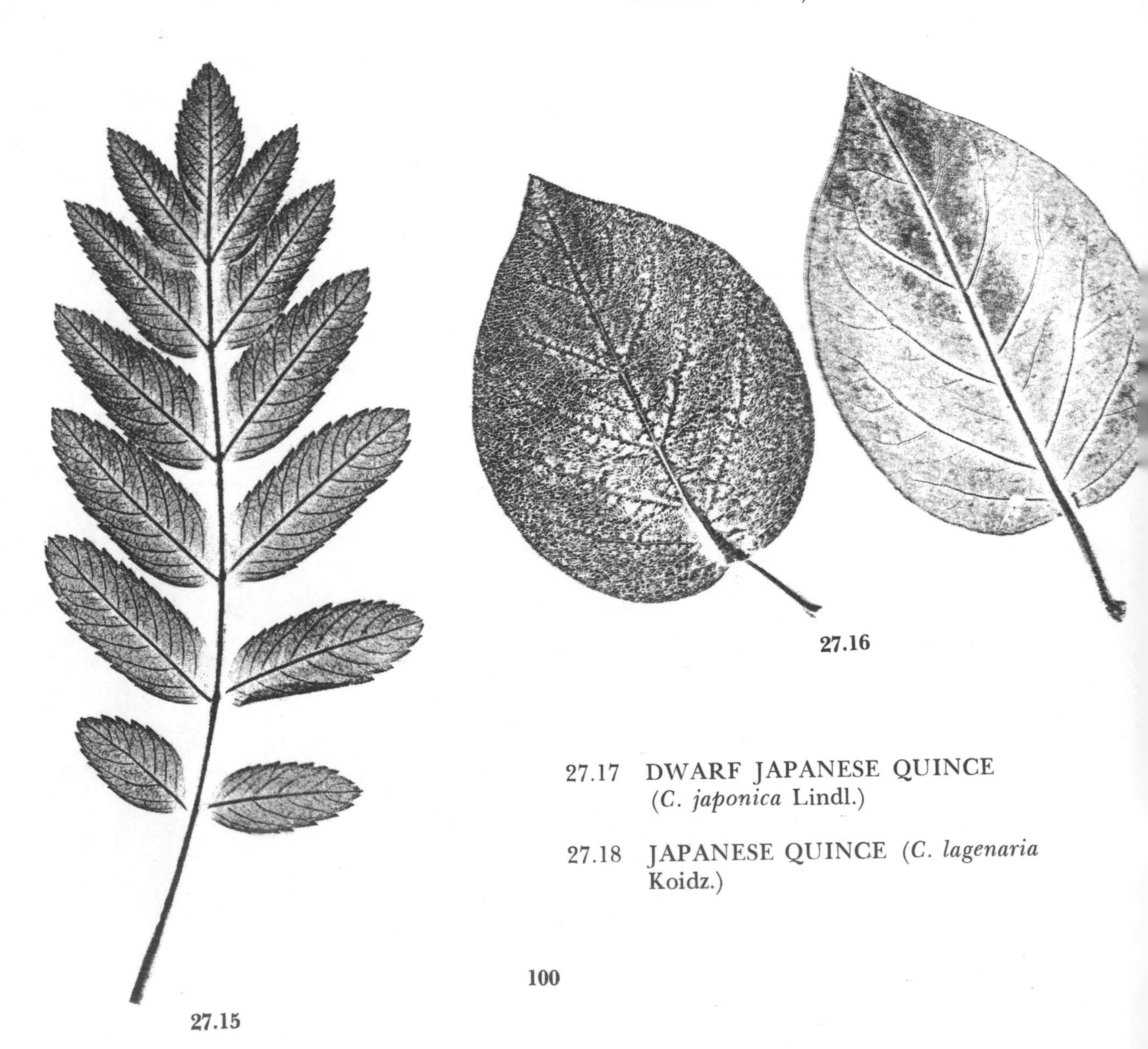

27.16

27.15

27.17 DWARF JAPANESE QUINCE (*C. japonica* Lindl.)

27.18 JAPANESE QUINCE (*C. lagenaria* Koidz.)

27.17

27.18

THE FIRETHORN GENUS
(*Pyracantha* Roem.)

This genus of about 6 species consists of ornamental woody plants from the temperate regions of the Old World. The fruits are small, red, decorative pomes persisting into the winter. It can be distinguished from the genus *Cotoneaster* by the presence of thorns.

27.19 FIRETHORN or HARDY PYRACANTHA (*Pyracantha coccinea* Roem.)

The fruits are edible, but not pleasant.

27.19

THE SERVICE-BERRY GENUS
(*Amelanchier* Medic.)

These shrubs and small trees are noted for their showy white flowers and edible dark-blue or black fruits. The ripe fruits are eaten raw, cooked, or dried, while the leaves are dried for tea. There are about 25 species in the genus.

27.20 SERVICE-BERRY; JUNEBERRY; SHADBUSH (*Amelanchier arborea* Fern.)

The berries of this species are edible, but rather tasteless. However, other species (e.g., *A. laevis, A. stolonifera,* and *A. canadensis*) bear delightfully sweet and juicy berries that are widely acclaimed. Nelson's Tree Nursery in DuBois, Pennsylvania specializes in the upbreeding and sale of these trees.

27.20

THE HAWTHORN GENUS (*Crataegus* L.)

There are about 1,000 species in this genus and one has to be an authority to be certain of an identification. They are hardwooded, and usually thorny, small trees. The fruits are small, red, yellow, or blackish pomes, which when ripe may be eaten raw, cooked, or dried. They are known to be eaten by 25 species of birds.

27.21 ENGLISH HAWTHORN (*Crataegus Oxyacantha* L.)

Twigs are green. Some varieties have red flowers.

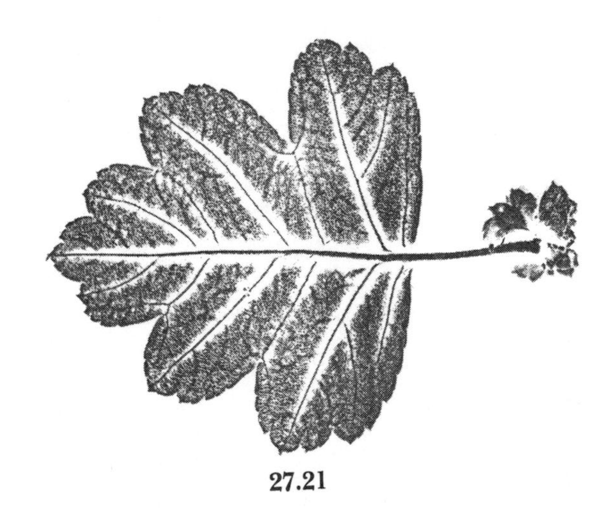

27.21

27.22 COCK'S-SPUR HAWTHORN (*C. Crus-galli* L.)

A bushy or dense shrub. Leaves are glossy, small.

27.22

27.23 SOFT, DOWNY, or MUSH HAWTHORN (*C. mollis* Scheele)

Leaves are large, downy. The fruit is edible, but watch for worms!

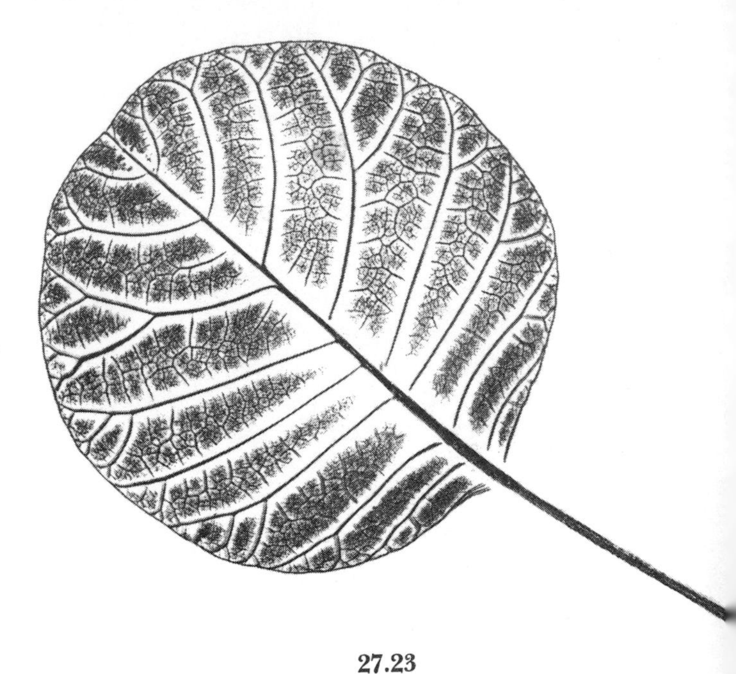

27.23

27.24 ONE-SEEDED HAWTHORN (*Crataegus monogyna* Jacq.)

27.24

27.25 DOTTED HAWTHORN (*C. punctata* Jacq.)

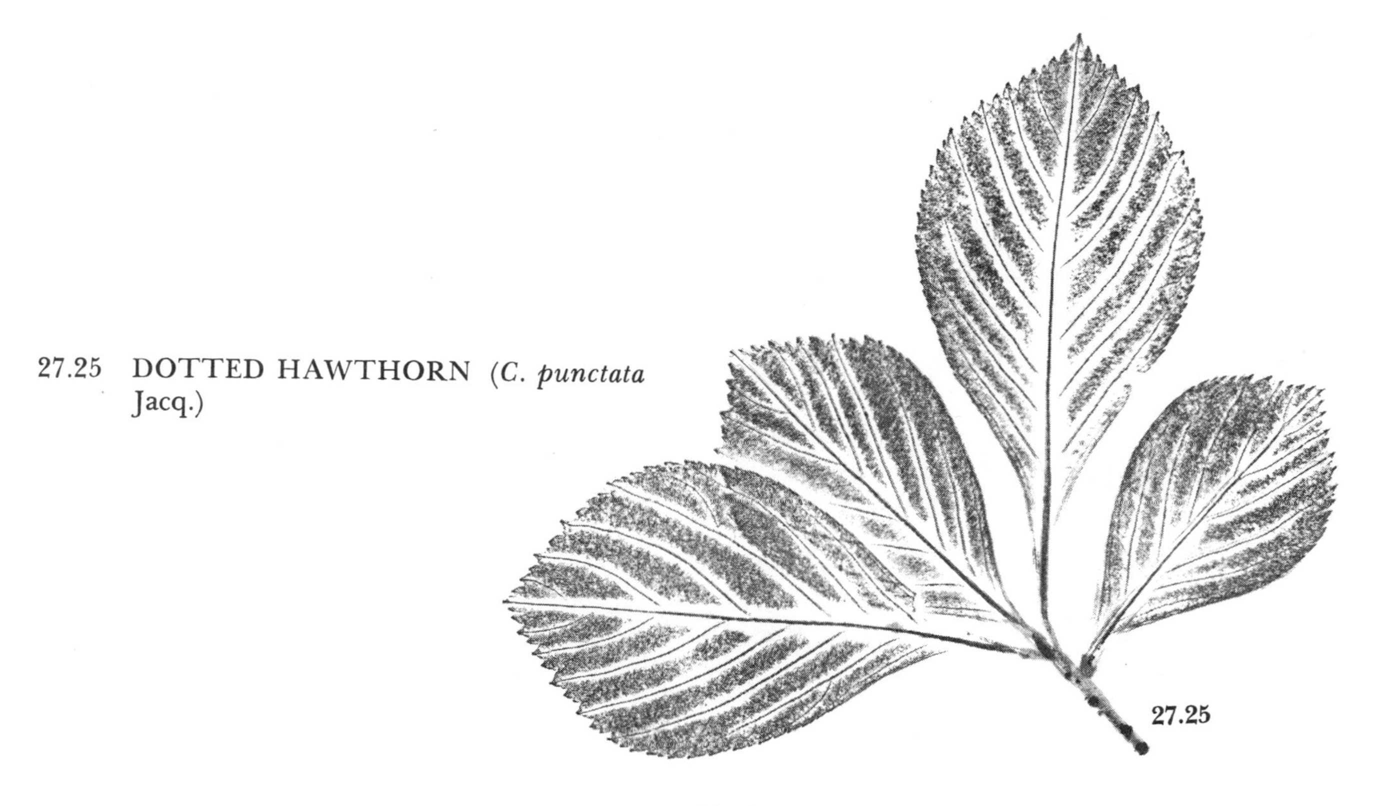
27.25

27.26 CARRIERE'S HAWTHORN (*C. Lavallei* Hérincq.)

27.26

27.27 WASHINGTON HAWTHORN (*Craetagus Phaenopyrum* Medic.)

This is one of many native trees now much better known in cultivation than as wildlings. It is, in fact, probably the most popular small tree. The berry-like fruits are produced in great abundance and are soft and palatable, giving this tree importance in wildlife management and as a survival plant. The seeds are soft enough to eat, but may contain a dangerous amount of hydrocyanic acid. The leaves turn bright red in the fall. The fruits of all Hawthorns were believed by the Eclectics to have important medicinal properties.

27.27

LEGUMINOSAE (FABACEAE)

Note: This family is subdivided into three subfamilies by some taxonomists in the following manner.

1. *Papilionoideae.* Those with bean or sweet-pea-like flowers.
2. *Mimosoideae.* Those with ball-like puffs of stamens as the most obvious part of the flower.
3. *Caesalpinoideae.* Those with open and more-or-less circular flowers.

The plants illustrated on the following pages are grouped to conform to this classification.

In speaking of the LEGUMINOSAE as a family, however, we can say that it is a large group of herbs, trees, and shrubs from many diverse habitats. And while they vary widely in form, as one might expect of some 13,000 species, they have one structure in common—a true pod or legume. This one-celled pod is usually dehiscent by one or both sutures into two parts, called valves. In some species, however, the pod is not dehiscent, and in others it breaks into joints.

The flowers of the *Papilionoideae* are typically the butterfly-like pea flowers with 5 sepals and 5 petals: One upright dorsal petal (the banner), two lateral horizontal petals (the wings), two lower ventral more-or-less united petals (the keel). The seeds, which develop in the pods, are exceptionally high in protein and hence the LEGUMINOSAE is one of the most important plant families serving mankind. Only nuts and a few other seeds that are difficult to crop annually in an agricultural operation contain as much protein.

Included in the family are many trees, a majority of which are from tropical regions. A few of these, such as TAMARIND and the CAROB, produce edible fruits. The seeds of LUPINS, WISTERIAS, and various others contain hydrocyanic acid and are very poisonous. If the occurrence of this dangerous acid has not been completely charted, this should be done, otherwise a great deal of high-protein food may continue to go begging and even a few unwarranted deaths may occur.

The legume crops have always been important because of their ability to put atmospheric nitrogen back into the soil. They effect this remarkable economy by means of the nitrogen-fixing bacteria that live symbiotically in nodules on the roots of the plant. In recent years agriculture in the U.S. has been less dependent on this natural process than heretofore because of the development of chemical means to achieve the same end.

THE YELLOW-WOOD GENUS
(*Cladrastis* Raf.)

This genus includes 4 species of deciduous trees of North America and Eastern Asia.

28.1 YELLOW-WOOD or VIRGILIA
(*Cladrastis lutea* (Koch)

The bark is gray. The dry pods are slightly constricted between the seeds.

28.1

About 20 species of ornamental trees with pea-like flowers.

28.2 SCHOLAR-TREE; PAGODA-TREE (*Sophora japonica* L.)

The undersides of the leaves are whitish. The pods, which contain a sticky pulp, are much constricted between the seeds.

28.2

About 20 species of American trees and shrubs whose stipules are often spines.

28.3 BLACK LOCUST (*Robinia Pseudoacacia* L.)

The leaves are fragrant when drying. The pods are red and green, becoming brown when mature. The wood is very heavy, strong, hard, and durable. The inner bark, leaves, and seeds contain poisons that cause vomiting, diarrhea, and cardiovascular changes. Do not confuse with HONEY LOCUST in the same family. BLACK LOCUST has two spines at the base of the leafstalk.

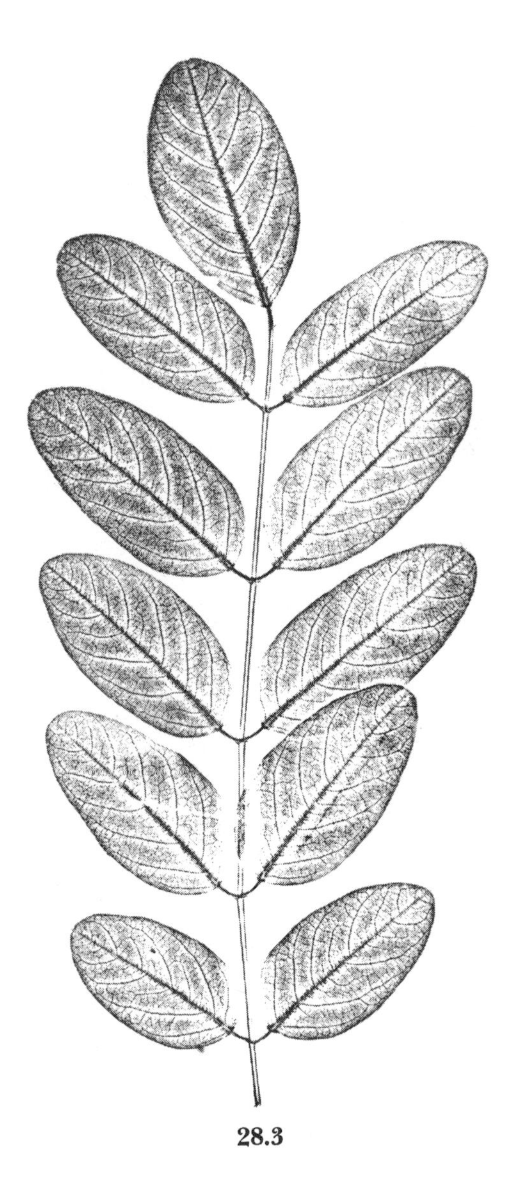

28.3

THE LABURNUM GENUS
(*Laburnum* Medic.)

All parts of the plants in the three species of this genus are poisonous. They are native in southern Europe and western Asia.

28.4 LABURNUM or GOLDEN-CHAIN TREE (*Laburnum anagyroides* Medic.)
Again note: The seeds and flowers are poisonous!

28.4

THE BAUHINIA GENUS (*Bauhinia* L.)

There are 150 to 200 species in this genus which, although native in the tropics, are widely planted in Florida and California. The leaves are divided by a cleft, sometimes so deeply as to make two distinct leaflets. The form suggests two brothers, hence the genus name, commemorating John and Caspar Bauhin, 16th-century herbalists, is quite appropriate.

28.5 BAUHINIA (*Bauhinia* sp., probably *corniculata*)

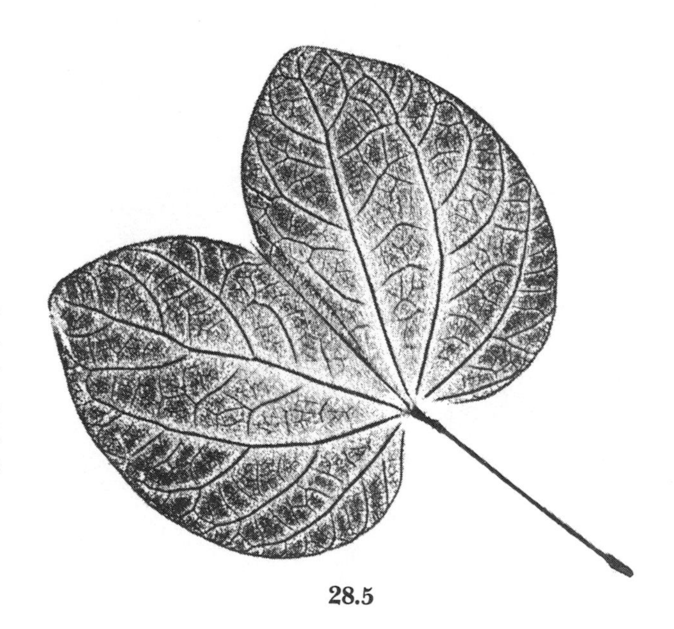

28.5

LEGUMINOSAE (FABACEAE)

THE ALBIZIA GENUS (*Albizia* Durazz.)

Some of the 50 species in this genus are valuable timber trees, and one species native in Mexico is planted for ornament in our Deep South. In this genus the pod is not constricted between the seeds.

28.6 HARDY-MIMOSA or SILK-TREE (*Albizia Julibrissin* Durazz.)

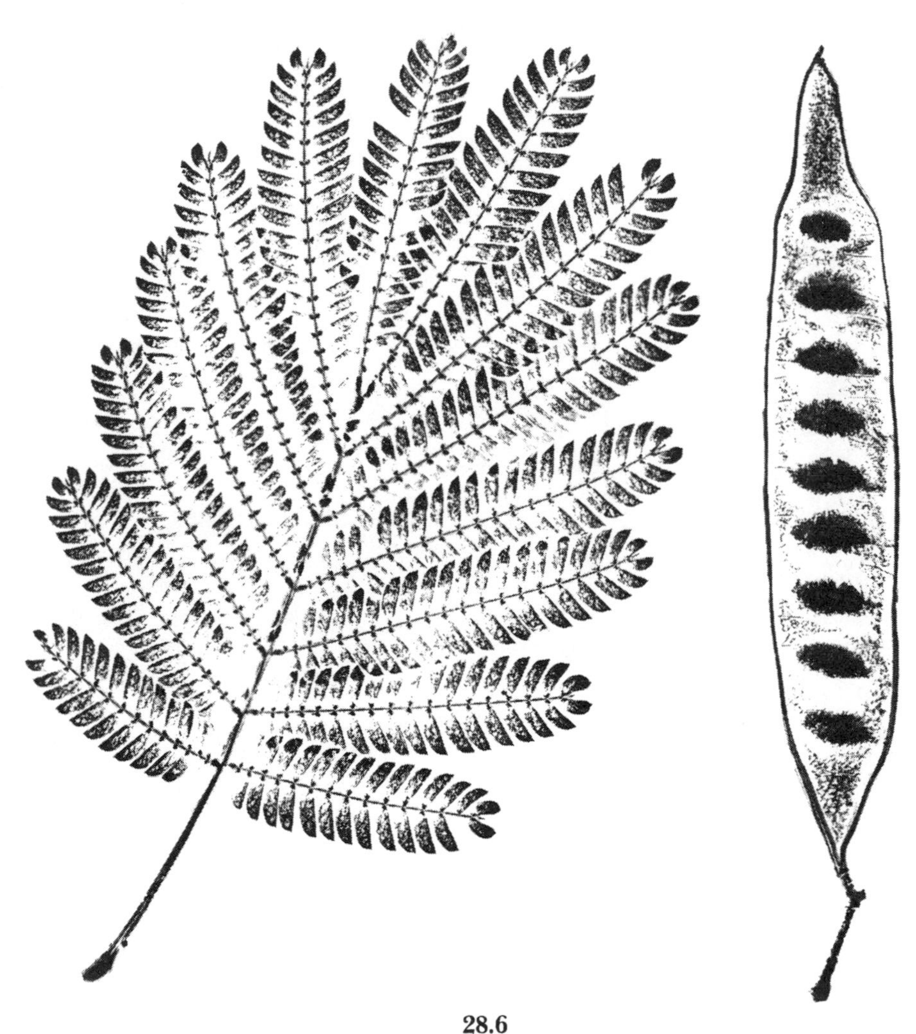

28.6

THE POINCIANA GENUS (*Poinciana* L.)

Large orange, red, or yellow flowers and finely-divided leaves are characteristic of the ornamental trees and shrubs in this genus.

28.7 DWARF POINCIANA (*Poinciana pulcherrima* L.)

THE HONEY LOCUST GENUS (*Gleditsia* L.)

About 12 species of this genus are distributed throughout the world.

28.8 HONEY LOCUST (*Gleditsia triacanthos* L.)

This species has 3- to 4-inch-long simple or branched spines on the trunk and larger branches. They are often three-pronged and have been used as pins and as spear-points. The edible pod-type fruits may feed cattle, deer, and rabbits.

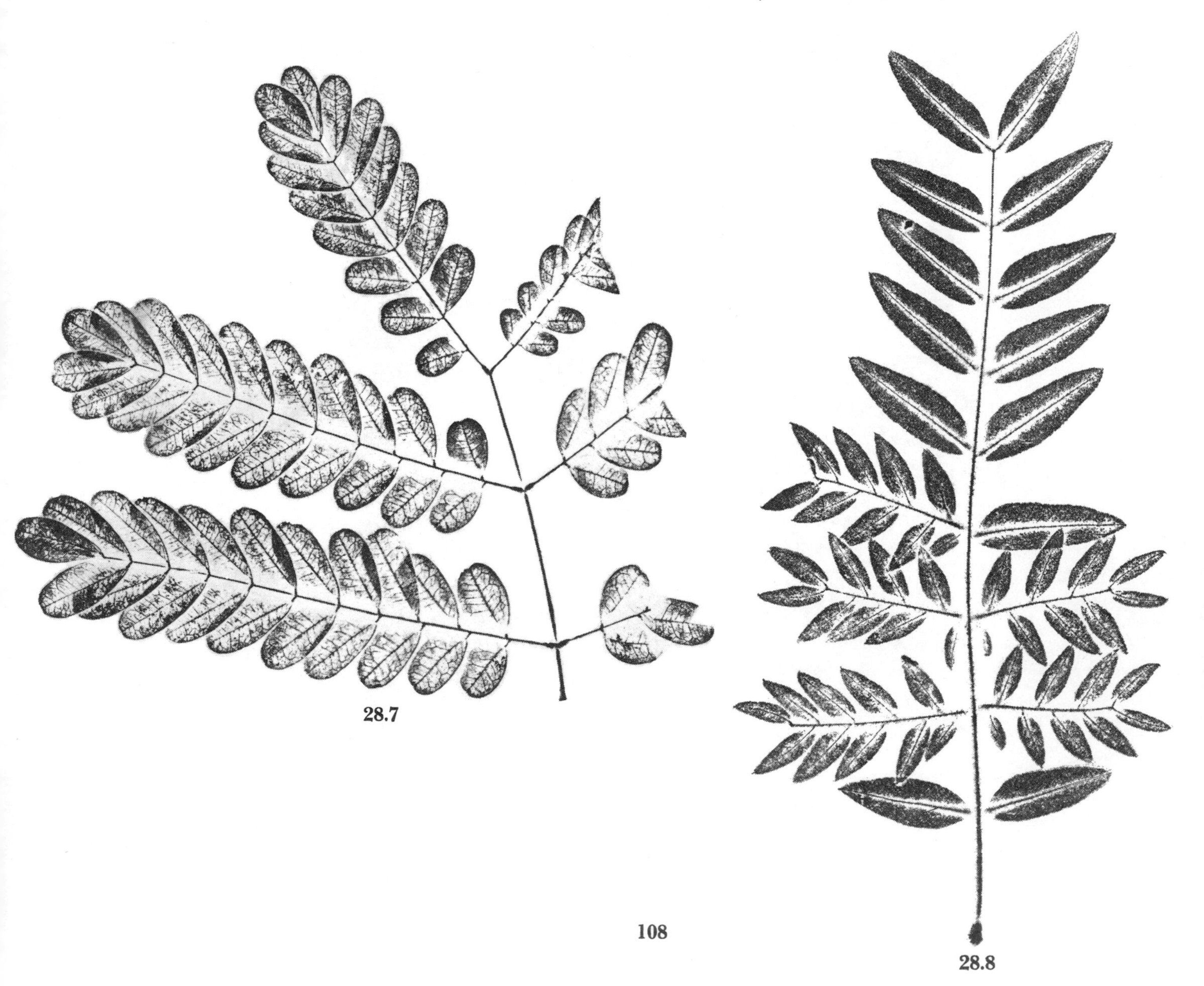

28.7

28.8

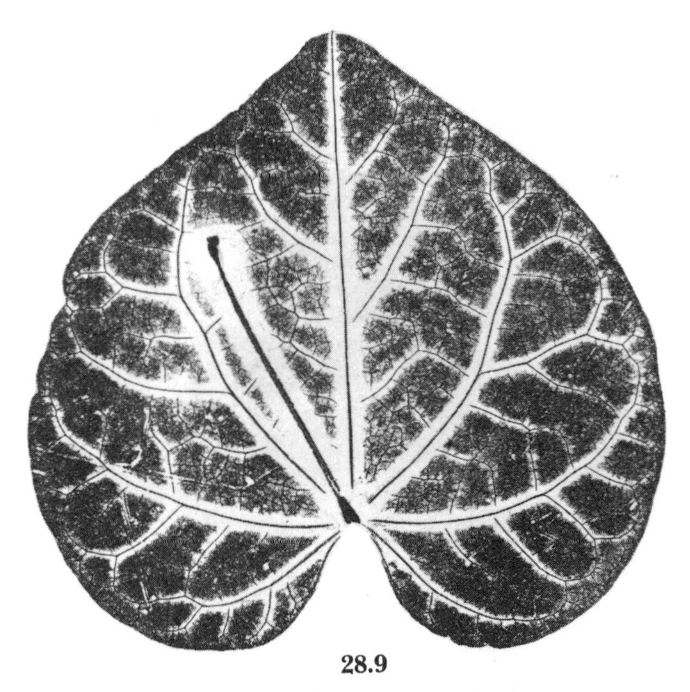

28.9

THE GYMNOCLADUS GENUS
(*Gymnocladus* L.)

Two species of deciduous trees native in North America and China. *Gymnocladus dioica* may reach 100 feet. Its fruits are reddish-brown pods up to 10 inches long, containing seeds that are approximately one inch long.

28.10 KENTUCKY COFFEE-TREE
(*Gymnocladus dioica* Koch)

In the past the seeds have been used as a substitute for coffee, but the seeds and the pulp of the pod that surrounds them contain a poison that, at worst, can bring on a coma. The tree is sometimes mistaken for the HONEY LOCUST, which is not poisonous.

THE REDBUD GENUS (*Cercis* L.)

Includes 7 species of ornamental deciduous trees and shrubs. The pink or red flowers occur in clusters or racemes; the leaves are palmately veined.

28.9 AMERICAN REDBUD or JUDAS-TREE
(*Cercis canadensis* L.)

Among the most popular native trees for ornamental planting. The leaves turn yellow in the fall but have a reddish or purplish sun-protective pigment (anthocyanin) when new leaves develop in bright sunlight. The colorful flowers are edible raw (in salads). The very young pods can be breaded and fried and have an agreeable sour taste. The seeds should be examined chemically and gastronomically. New bark with reddish color will dye red. The pliant shoots are used in basketry. The leaves are natural valentines (at the wrong time of the year).

28.10

There are both small and large ornamental and commercially valuable trees in this family of more than 100 genera and over 1,000 species. They are to be found growing wild in many parts of the world. The characteristics that they share in common are: aromatic leaves, the elaboration of essential oils, and edible fruits.

THE CITRUS GENUS (*Citrus* L.)

This is undoubtedly one of the more important genera in the plant kingdom. The fruits not only keep many people healthy with their citric acid and vitamin C, but the citrus industry also furnishes a livelihood for thousands of growers and workers in California, Texas, Florida, and the Mediterranean countries.

Bottling, canning, and freezing have revolutionized this industry as many another. Diverse associated products are made from the processed citrus fruits. Pectin, used for jelling, is recovered from the pulp, and the essential oils, used for flavoring and in perfumery, are extracted from the skin. But the citric acid, also called lemon salt, originally a by-product of fruit processing, now comes to us almost entirely through a process of bacterial fermentation.

Citrus fruit trees are more easily identified by their fruits than by their leaves. What appear to be "simple" leaves are actually modified compound leaves in which only one leaflet remains attached to the end of a jointed midrib. The amount of winging to be found on the so-called petioles is an important aid in identifying the species if the fruit is not present.

The leaves of all species are agreeably aromatic.

29.1 SWEET ORANGE (*Citrus sinensis* Osbeck.)

29.2 LEMON (*C. Limon* Burm.)

29.1

29.2

29.3 GRAPEFRUIT (*Citrus paradisi* Macf.)

29.4 LIME (*C. aurantifolia* Swingle)

29.5 CALAMONDIN (*Citrus mitis* Blanco)

A small tree or tub plant grown chiefly for ornament but the fruits can be used for ades and marmalade. There is little or no winging of the petiole. (Considered to be a hybrid with no authentic Latin name.)

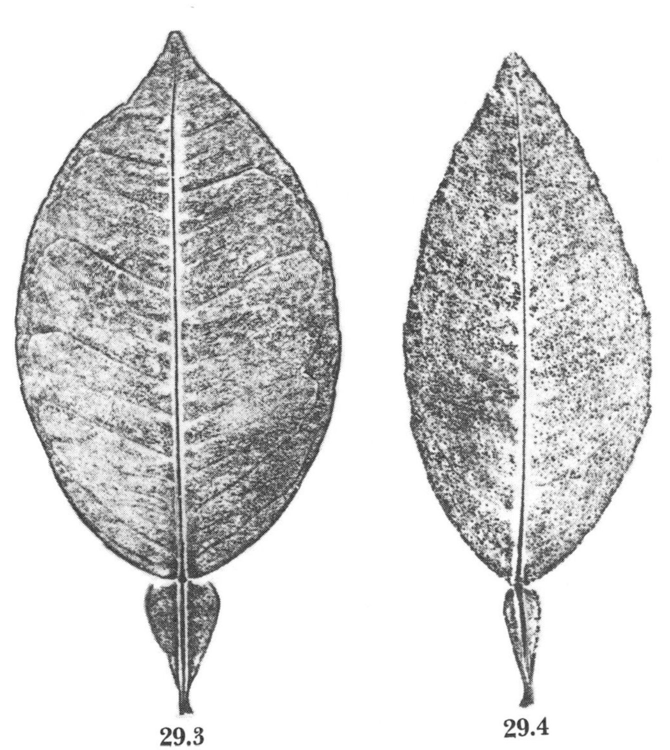

29.3 29.4

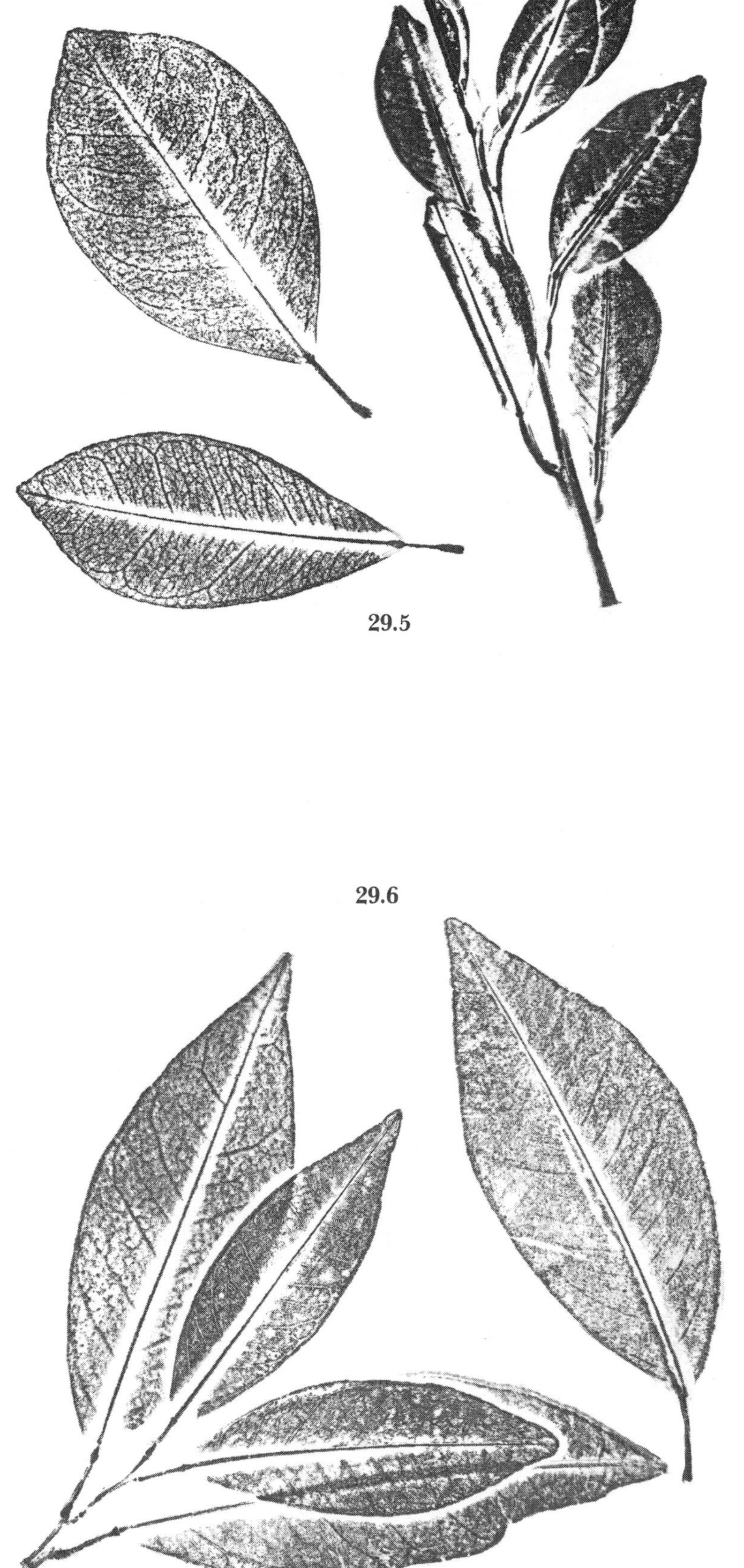

29.5

29.6

THE KUMQUAT GENUS
(*Fortunella* Swingle)

There are 4 or 5 species of East Asian trees in this genus whose fruits differ from that of a Citrus in having only 3 to 6 segments and in being less than 1½ inches in diameter.

29.6 OVAL and NAGAMI KUMQUAT
(*Fortunella margarita* Swingle)
A small tree occasionally planted in the subtropical parts of Florida. The petiole is only slightly winged. The juice is sour, the pulp neutral. Its fruits are preserved whole or eaten with sugar.

OTHER RUTACEAE

BAEL-TREE (*Aegle Marmelos*)

Highly prized for its fruit in the Orient but it is rarely seen on this continent. The fruit is reputed healthful and medicinal.

WHITE-SAPOTE (*Casimiroa edulis*)

Occasionally planted and thriving at Miami. The leaves are palmately compound in the manner of a BUCKEYE. The fruit is very soft, entirely unlike a citrus fruit, with luscious flesh and thin, usually bitter, skin.

LIMEBERRY-BUSH (*Triphasia trifolia*)

Rather common around Miami. It is dense-growing, spiny, and makes a superior hedge. The leaves are similar to those of the TRIFOLIATE-ORANGE but usually smaller. The small red fruits are almost too tart and turpentiny to eat raw, but are favored by some, with plenty of sugar, for preserves. I have often thought that they would make a colorful addition to the thousands of fruit baskets annually shipped out of South Florida.

THE PONCIRUS GENUS (*Poncirus* Ref.)

A genus containing only one species of very thorny small trees native in China and more hardy than the Citrus.

29.7 TRIFOLIATE-ORANGE (*Poncirus trifoliata* Raf.)

A large deciduous shrub hardy as far north as Cincinnati and New York City. Its small, dryish fruits can be used for ades but are far inferior to the true CITRUS. It has been used for hybridizing with, or for serving as scion stock for, the more tender CITRUSES. As is usual in this family, the flowers are fragrant and the leaves aromatic when crushed.

THE MURRAYA GENUS (*Murraya* L.)

The flowers are fragrant in these approximately 11 species of spineless small trees and shrubs.

29.8 ORANGE-JESSAMINE (*Murraya paniculata* Jack)

One of the favorite hedge and specimen shrubs around Miami where anything, related or not, with white fragrant flowers is a "Jasmine." The small red berries are acrid and inedible.

CURRY-TREE (*M. Koenigii*)

Its aromatic leaves are an ingredient of the Oriental seasoning called curry.

THE ZANTHOXYLUM GENUS (*Zanthoxylum* L.)

About 150 species of widely distributed, prickly, aromatic trees and shrubs are included in this genus.

29.9 PRICKLY-ASH or TOOTHACHE-TREE (*Zanthoxylum americanum* Mill.)

In folk remedies the twigs are given as a source of drugs that are to be chewed to stop toothache. But these are hard and thorny, while the equally useful soft leaves are not often mentioned. All parts are remarkably sialogogue, inducing a strong flow of saliva. Tehon says the berries are bitter, stimulant, and diaphoretic. These red berries open to expose the shiny black seeds, the two colors most attractive to disseminating birds.

29.10 TOOTHACHE-TREE, HERCULES'-CLUB (*Zanthoxylum Clava-Herculis* L.)

Also used to break a fever and induce sweating.

29.7
29.9
29.8
29.10

THE CORK TREE GENUS
(*Phellodendron* Rupr.)

In Eastern Asia there are about 9 species of these closely related ornamental trees.

29.11 AMURLAND CORK-TREE
(*Phellodendron amurense* Rupr.)
A large hardy tree with soft corky bark, black berries, and aromatic (turpentine) leaves. The corky bark is aesthetically pleasing. A beautiful specimen is to be seen at St. Bartholomew's Church, Park Avenue, New York City.

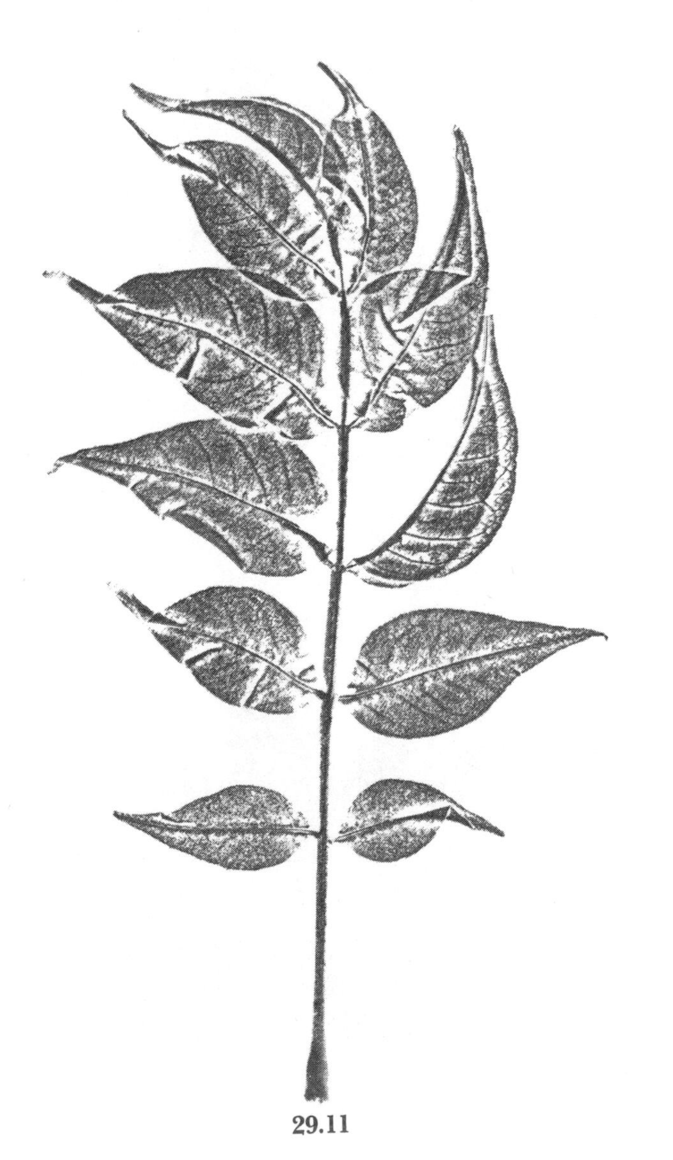

29.11

29.12

THE HOP TREE GENUS (*Ptelea* L.)

An ornamental North American genus of shrubs and small trees consisting of about 10 species.

29.12 WAFER-ASH or HOP-TREE (*Ptelea trifoliata* L.)
A large native shrub occasionally planted for ornament. The flat "fruits" are used in winter bouquets. They contain an agreeable-tasting oil but are difficult to tear open. The bark of the roots is tonic, astringent, and sedative, of value because nonirritating. It is also an important winter food of rabbits. The use of the fruits in brewing beer is a bit of folklore passed down from one book to another.

There are some 150 species in the 30 genera that comprise this family. A unifying characteristic of this group of trees and shrubs is their bitter bark.

THE AILANTHUS GENUS
(*Ailanthus* Desf.)

There are perhaps 10 species in the genus, of which only one is seen planted as a shade or ornamental tree. The fruits consist of a group of 1 to 6 two-inch-long oblong samaras, each with a flattened seed in the middle. Glands are located along the margins on the reverse side of the leaf (note the arrow on the leaf-print).

30.1 TREE-OF-HEAVEN (*Ailanthus altissima* Swingle)

When I was a boy we dubbed it STINKWOOD. We made spear-throwers of the thick twigs (see Jaeger on atlatls). The streaked bark, showy red "fruits," pith of twigs, and other parts are excellent for craft-work. The fetid leaves and the bark are poisonous, also medicinal, with an action on the brain, but usually avoided by the medical profession. The flowers of the male trees send out a powerful fetor and the roots are accused of clogging pipes. This species will grow in any crack in the pavement.

30.1

This is a widely distributed family of some 250 genera and 8,000 species of herbs, shrubs, and trees that have ornamental, medicinal, and, in some cases, food value. The juice is often milky.

THE BREYNIA GENUS (*Breynia* Forst.)

From Asia, Africa, and the islands of the Pacific come some 15 species of trees and shrubs.

31.1 SNOW-BUSH (*Breynia nivosa* Small)
May be seen planted as a hedge in the South or in greenhouses in the North. Has dark-red zig-zag branches. The fruit is a berry. It now runs wild in Florida.

31.1

THE CROTON GENUS (*Codiaeum* Juss.)

There are 6 species of this genus growing wild in Malaya and the Pacific Islands, but only one is commonly seen here in the U.S.

31.2 CROTON (*Codiaeum variegatum* Blume var. *pictum* Muell. Arg.)
A shrub or small tree with leaves variously marked with white, yellow, or red. Also in many cultivated variations with leaves lobed, cut, crisped, and so on.

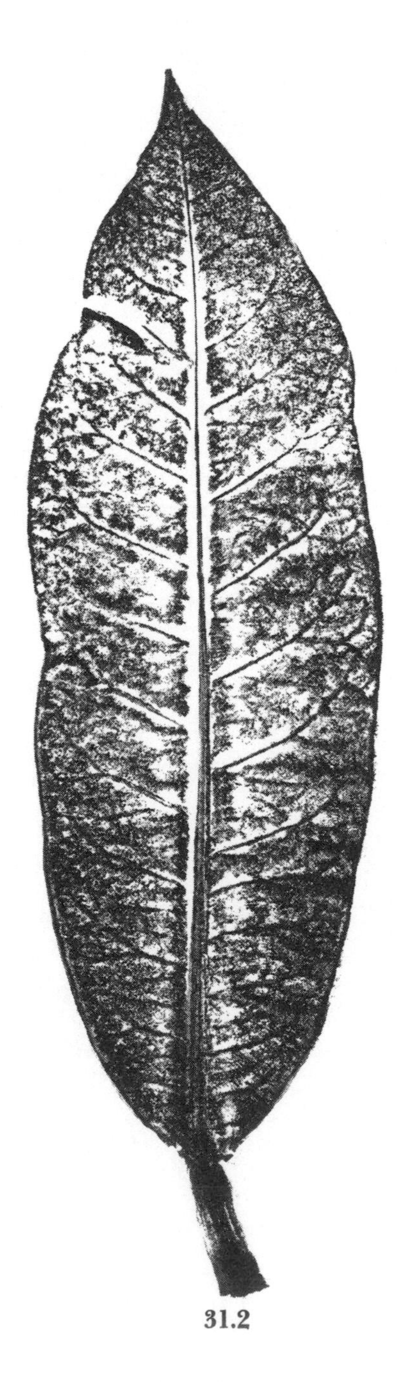

31.2

THE ALEURITES GENUS
(*Aleurites* Forst.)

This genus includes four species of oil-producing and ornamental trees of Eastern Asia and the islands of the Pacific.

31.3 CANDLENUT; VARNISH-TREE
(*Aleurites moluccana* Willd.)

This tree may reach 60 feet in height. It has a four-angled fleshy fruit with large, rough, poisonous seeds. In the Philippines the oil is called lumbang oil, and it is used as a drying oil in paints, varnishes, lacquers, and linoleum. From its use as an illuminant comes the name CANDLENUT.

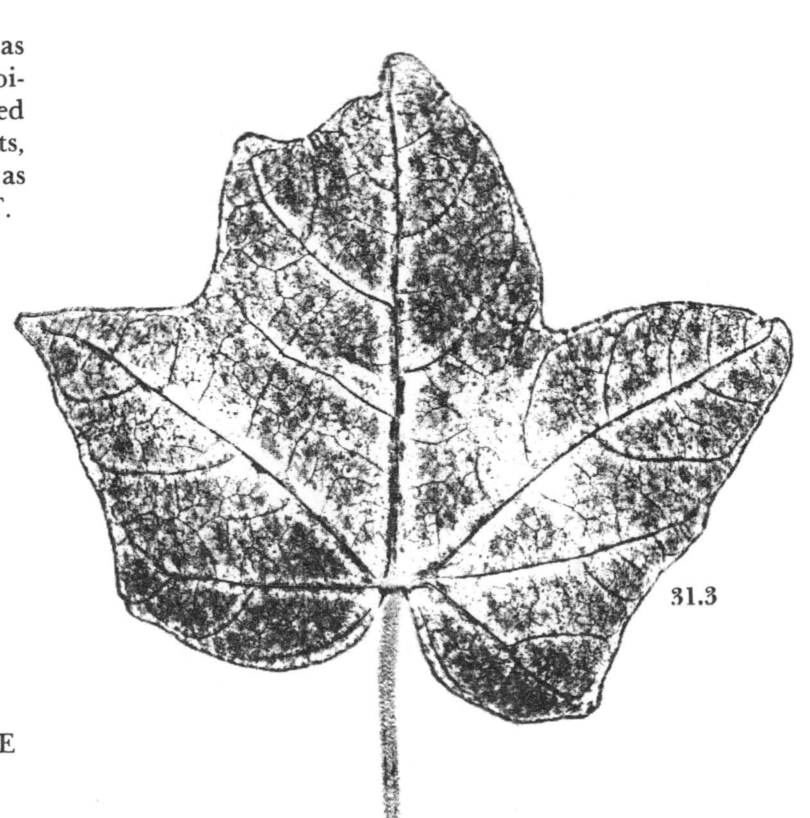

31.3

OTHER EUPHORBIACEAE OF NOTE

POINSETTIA (*Euphorbia pulcherrima* Willd.)

A small shrub with floral leaves of brilliant red, pink, or white, which blooms in mid-winter and hence is widely used at Christmas-time.

CASSAVA (*Manihot esculenta* Crantz.)

A tropical root crop that furnishes the basic food for millions of people. A slight amount of heat drives off a poisonous glucoside and the starch that remains may be eaten raw or cooked. Cassava bread, an intoxicating beverage, a healing medicinal, and tapioca are among the diverse by-products of this plant.

In the approximately 60 genera and 400 species that comprise this family, there are trees, shrubs, and woody vines but apparently no herbaceous plants. In many instances, the "fruit" is an outstanding feature of the plant because it has relatively large seeds and thin pulp. This is true from the tiny SUMAC bobs to the large MANGO fruits. The family also includes the two important nut trees, CASHEW and PISTACHIO, in which the seed coat is hard and dry.

The *Cashew* family is also one of the more consistent producers of tannin and tannic acid and its variations, such as gallic acid. These chemicals are only slightly acidic, but they are notable for their astringency. They are useful in medicine for keeping burns from suppurating, as astringents, and for other purposes. They also have numerous industrial applications.

Tannic acid has the property of precipitating proteins and the "tanning" of leather is based on this reaction between the tannin and the proteins in the hide. Some leathers are cured with metal salts but most shoe leathers with tannins.

It would be interesting to know the adaptive value of tannins to the trees and shrubs that produce them. Some work suggests that they are a part of the oxidative mechanism of the plant and others indicate a value in wound healing. Unfortunately, there has not been enough observation or experiment along this line. Note that the SUMAC tannin comes from leaves and galls, CHESTNUT tannin from the wood, HEMLOCK-TREE tannin from the bark, and OAK tannin chiefly from the bark, but also from galls—not to mention BLACKBERRY and WILD GERANIUM tannin, which comes from the roots or other underground parts.

POISON IVY and POISON SUMAC, two members of the family that cause dermatitis, are considered below. All the others are aromatic, innocuous plants and, except for CASHEW and MANGO, of no great economic importance—even though, in addition to tannin, many yield fruits, nuts, resins, and drugs.

THE MANGO GENUS (*Mangifera* L.)

There are about 30 species in the genus, but the only one widely grown is *M. indica,* a large tree noted for its edible fleshy drupes containing a compressed fibrous stone—the MANGO of commerce.

32.1 TRUE MANGO (*Mangifera indica* L.)

This tree is much planted in South Florida and its delicious fruits are not rare in northern markets. Their season is June, unless they are imported from farther south. Persons allergic to it can get a rash from eating the fruit of this tree. It is the skin that does the damage and peeling obviates this. The fruit can also be cooked. The Haden Mango was developed by David Fairchild.

VI-APPLE, AMBARELLA-TREE (*Spondias cytherea* Sonn.)

Next to the MANGO, probably the most promising Anacardiaceous fruit in Florida. The fruits are most enjoyable, juicy even though not so luscious as MANGOS. An objection to them is the spicule-studded core extending into the flesh. The tree has a spreading habit and is very handsome when bearing its yellow fruits. It freezes back in cold winters but regenerates strongly the next year.

THE COTINUS GENUS (*Cotinus* Adans.)

The genus consists of only 2 species of shrubs and small trees whose flowers occur in large panicles. The foliage is quite ornamental. This genus is a segregate from the SUMACS and was originally classified under *Rhus,* but the leaves and other parts are quite different.

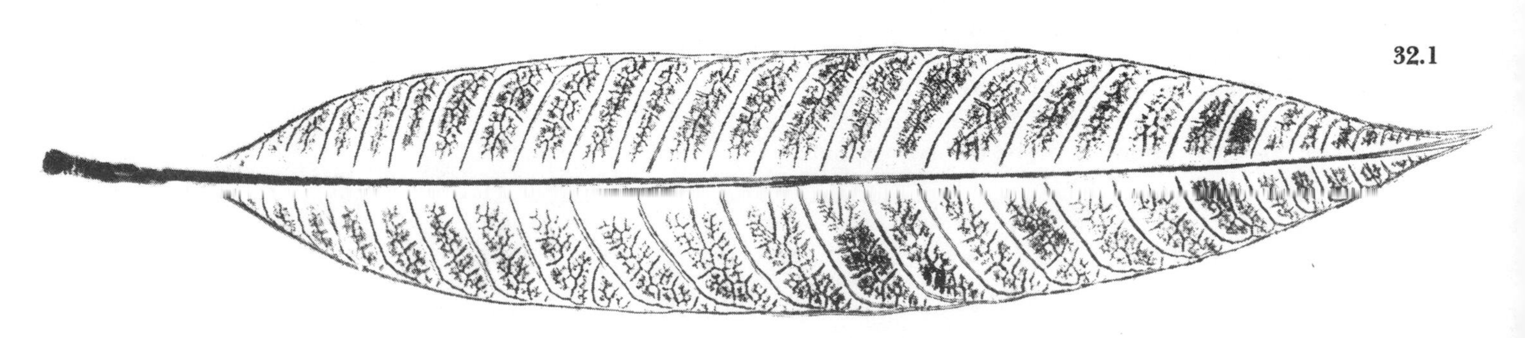

32.1

32.2 EUROPEAN SMOKE-TREE (*Cotinus Coggygria* Scop.)

A large rotate shrub, or sometimes tree-like. The leaf blade is nearly spherical, the whole leaf suggesting a ping-pong bat. In one variety the leaves are purple through the year. The most conspicuous feature of this shrub is the feathering on the seed stalks, very valuable for craft-work and decoration. To the plant itself this feathering fosters seed dispersal either by the wind or by attracting birds. The leaves are aromatic and medicinal.

32.2

AMERICAN SMOKE-TREE (*C. americanus* Nutt.)

This is native in the Smoky Mountains and is sometimes cultivated, although not so much as it merits to be. In keeping with our sunnier falls, it produces high fall leaf coloring for attracting disseminating birds and neglects the fuzzing of the stems resorted to by the European species, which grows where muggy falls do not induce fall coloration.

THE PEPPER-TREE GENUS (*Schinus* L.)

About 15 species of resinous trees make up this genus. Two are grown rather commonly in Florida and California as ornamentals. (The *ch* of *Schinus* is pronounced *k* as in chemistry.)

32.3 BRAZILIAN PEPPER-TREE; CHRISTMAS-BERRY-TREE (*Schinus terebinthifolius* Reddi.)

Also locally called FLORIDA HOLLY. A SUMAC-like bush; when it grows to tree stature and is unmolested it forms bushlets at the ends of the spreading branches. The leaves smell like turpentine when crushed (whence the specific epithet).

The berries are produced in abundance and are used in winter decoration. Through distribution by birds this shrub-tree has taken over much of the waste land near Miami.

32.3

CALIFORNIA PEPPER-TREE (*Schinus Molle* L.)

This species is quite different from the one above, being of graceful rather than of stiff growth. It is much planted in California, and only rarely in Florida, but whether this is a matter of climate or of custom, I do not know.

The name *Schinus* is the ancient name of the MASTIC TREE (*Pistacia Lentiscus*) of the Old World. Adding the suffix *opsis,* meaning "of the appearance of," to *Schinus* gives the name of the South American genus *Schinopsis* containing *S. Laurentzii* and *S. Balansae,* now by far the most important sources of tannin in the world. Their common name QUEBRACHO comes from "madera que bracha" or wood-that-breaks, for axes and machetes fare poorly with it.

THE SUMAC GENUS (*Rhus* L.)

The approximately 150 species of this genus consist of trees and shrubs that are characterized by their milky or resinous juice. Some are very poisonous to the touch; others are grown as ornamentals.

In addition to our American SUMACS, which are the ones we see daily, several Old World species yield lacquer, dyes, tannin, and drugs. Our native SUMACS would bear unlimited chemurgic exploitation, and this is especially true of SMOOTH SUMAC. Oil could be obtained from the seeds, acids from the berry pulp, paper and plastics from the wood. The abundant pith has craft uses and might be made into moxas.

32.4 SMOOTH SUMAC (*Rhus glabra* L.)

This native shrub grows abundantly and exuberantly in very poor soil where few other plants will thrive. It increases rapidly by underground runners as well as from seed, and would be very easy to propagate. It is capable of producing a multiplicity of products—tar, oils, acids, paper, plastics, wax, perfume, varnish, drugs, and, doubtless, many more; in addition to this, it has craft and landscape value. When this shrub grows along railroads where it is cut to the ground annually—conditions that could be reproduced in cultivation—it develops long, pithy shoots that can be made into floats, rafts, and many other camp articles. Being a native American and not a native of Europe, this shrub has been neglected. It offers a challenge and a promise.

32.5 STAGHORN SUMAC (*Rhus typhina* Torner)

Most of the SUMACS and, particularly, some western American species are called LEMONADE-BUSH, because beverages and a sort of vinegar can be made from the bobs. This species is better for this purpose than the SMOOTH SUMAC alongside because the bobs are coated with longish hairs instead of being merely roughened. The leaves of both species, but especially the SMOOTH, are collected in parts of the country and sold to tanners and drug wholesalers.

32.6 WINGLEAF or SHINY or DWARF SUMAC (*Rhus copallina* L.)

A very ornamental shrub turning brilliant red in the fall. The winged midrib of the leaf is distinctive. The berries are purplish in color. The yellow wood is used as a dye—called "young fustic." The specific epithet implies that the resin derived from this species could be used for varnish.

32.7 FRAGRANT SUMAC (*Rhus aromatica* Ait.)

This attractive shrub has two important uses—one medicinal and the other horticultural. It has long been an important botanical drug. Low forms of it are used at the Arnold Arboretum to "tie tall shrubs and trees to the ground." It is much planted at Cincinnati for holding the new superhighway banks. In fall the veins color before the rest of the leaf. The berries are so light in weight that they may be transported by the wind as well as by birds.

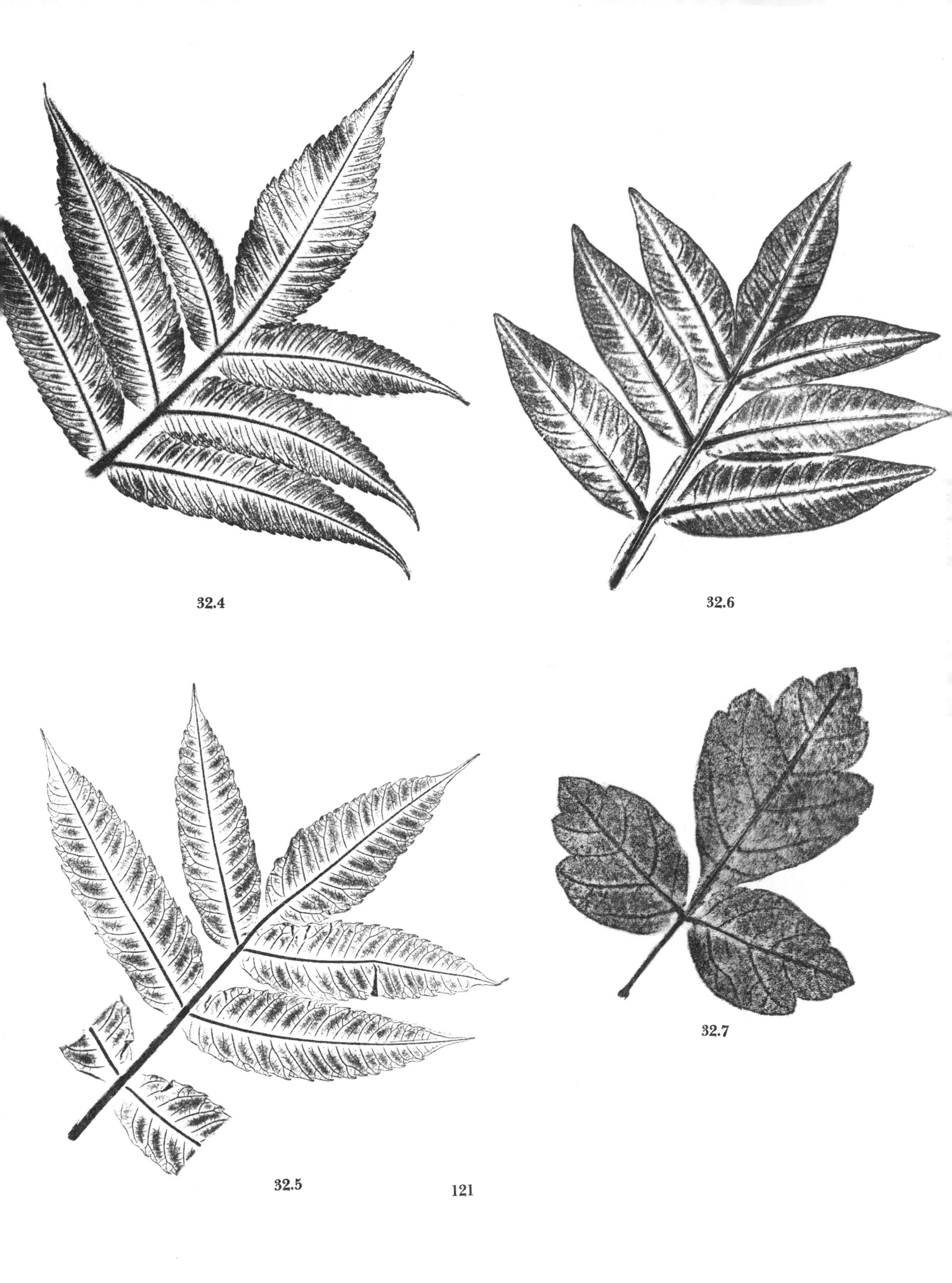

32.4

32.6

32.5

32.7

32.8

32.8 POISON-IVY (*Rhus radicans* L.)

Doubtless this vine is the commonest and most-feared of all touch-poisons. Many antidotes have been proposed against it, but apparently all they can do is relieve the symptoms. The poisonous principle is called urushiol and it appears to be a nonvolatile oil that can be spread by pets, smoke, tools, etc. In 1884 Laurence Johnson said that the principle is volatile. Dr. Johnson was wrong in this, but so right in warning foolhardy demonstrators against swallowing this poison! Nonetheless, the berries are eaten by over 50 species of birds, and the undigested seeds are spread by them.

32.9 POISON SUMAC (*Rhus Vernix* L.)

If anything, the poison from this shrub is even more virulent than that of POISON-IVY. And, as with the other, the brilliant fall colors invite trouble. Note that in both these species the berries are whitish or creamy-yellow, while in the non-poisonous species they are reddish. I have seen this shrub in a botanic garden where it was placed on an island, and in the rows of a nursery just somewhat out of the way. It might be better to leave this public enemy in its native swamps, where the people who see it are more likely to recognize it.

32.9

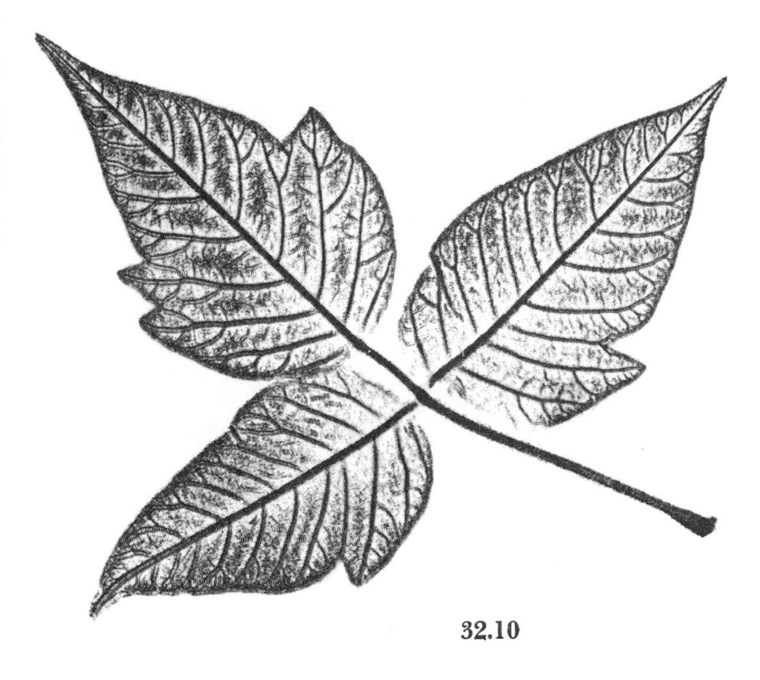

32.10

POISON IVY!

32.10 POISON IVY (*Rhus radicans* L.)

Print 32.8 illustrates POISON IVY in the spring, when its poison is most virulent. The swatch above is of a single leaf, to help you recognize it in the summertime. I once overheard an Indian "Princess" explaining to her non-Indian husband that the plants he was uprooting were definitely not Poison-Ivy and he had nothing to fear. They were, and he had!

OTHER ANACARDIACEAE

POISON-OAK (*Rhus Toxicodendron* L.)

Forms of POISON-IVY with lobed leaves, and often POISON-IVY in general, are miscalled POISON-OAK. Also some books speak of POISON-OAK as having a westerly range, implying that the ranges of the two species do not overlap. They do. The POISON-OAK leaf is hairy on both sides.

POISONWOOD (*Metopium toxiferum*)

A common native shrub of South Florida. It is at least as poisonous to the touch as the other species discussed. It is quite ornamental, with bright orange-colored berries and glossy leaves, and it is sometimes inadvertently allowed to grow on properties.

CHINESE SUMAC (*Rhus verniciflua*)

Not itself seen in this country, woodenware coated with its lacquer has caused dermatitis.

BATOKA-PLUM (*Pleiogynum Solandri*)

A large and comely tree deserving of further planting. Some are to be seen along the streets and in the City Cemetery of Miami. Its purple fruits are edible. The seed is so sculptured as to resemble a large brachiopod, interesting for craft-work.

The two economic species of Pistacia, *P. vera,* which produces the PISTACHIO-NUT of commerce and *P. Lentiscus* which produces MASTIC, an important source of varnish, are seldom, if ever, grown in this country. The species found in greenhouses is *P. chilensis,* a "mere" ornamental. PISTACHIO-NUTS are the chief flavoring agent of the superb green ice cream known as spumoni.

AQUIFOLIACEAE

Although centered in Central and South America, the *Holly* family, of 3 genera and about 300 species, is widely distributed. These trees and shrubs are mostly ornamental, but some species have been used medicinally in the past. The persistent red "berries" (one-seeded pyrenes) are a popular Christmastide decoration. Some 22 species of birds have been recorded feeding on these berries.

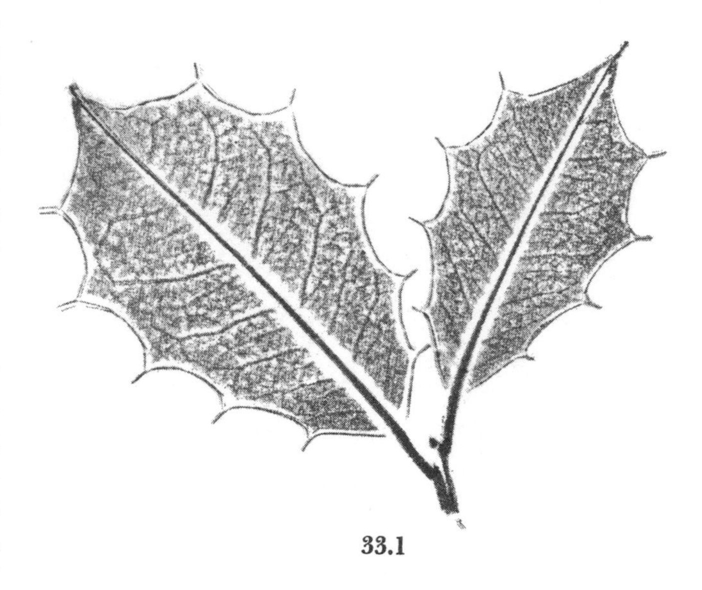

33.1

33.1 AMERICAN HOLLY (*Ilex opaca* Ait.)
A native tree of our east coast and widely planted. The male and female flowers are on separate trees. The berries contrast beautifully against the spiny leaves, and while they are bitter and inedible, they are not poisonous. In pioneer times the inner bark was fermented to make birdlime, a substance remaining sticky on exposure to air and so used to entangle small birds.

33.2 ENGLISH HOLLY (*Ilex Aquifolium* L.)

33.3 WINTERBERRY HOLLY (*Ilex verticillata* Gray)

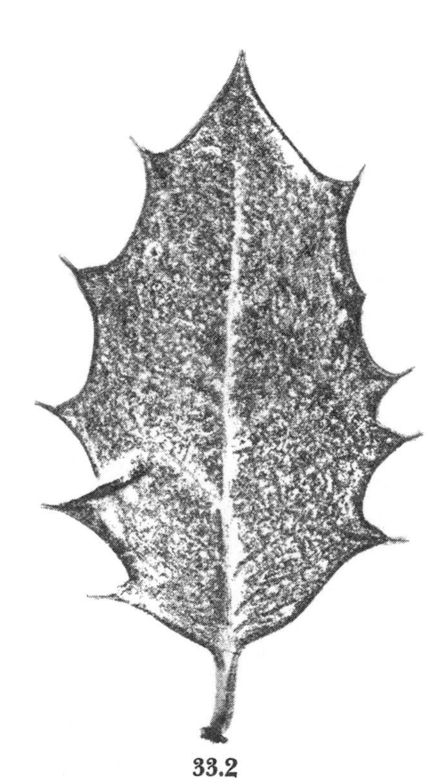

33.2

33.3

AQUIFOLIACEAE

33.4 INKBERRY or GALLBERRY HOLLY (*Ilex glabra* Gray)
The fruit is black.

33.5 POSSUM-HAW HOLLY (*Ilex decidua* Walt.)

33.6 YAUPON HOLLY (*Ilex vomitoria* Ait.)
Long known and probably used for the purpose suggested by its specific name. (Also the source of a harmless drink made from the leaves under the name of "Yaupon tea.")

33.7 PARAGUAY-TEA; YERBA MATÉ (*I. paraquariensis* St. Hil.)
This and perhaps other close relatives are sources of caffeine beverages known as maté.

33.4

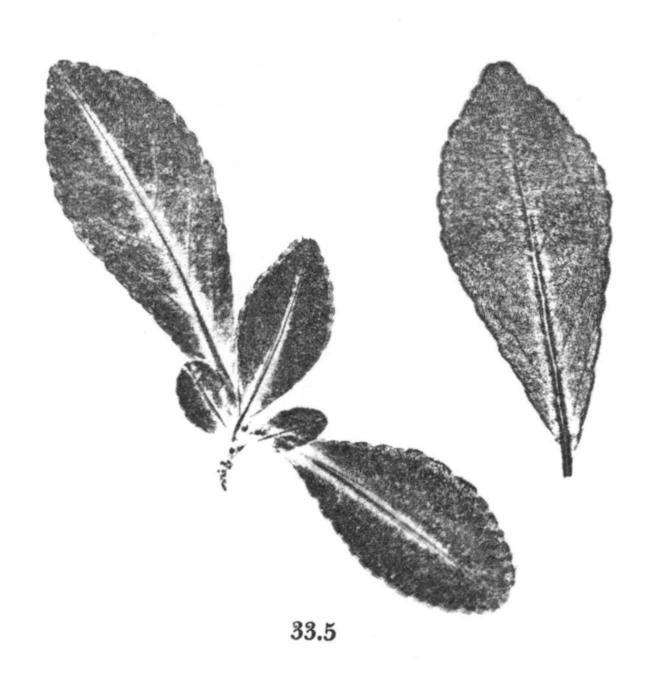

33.5

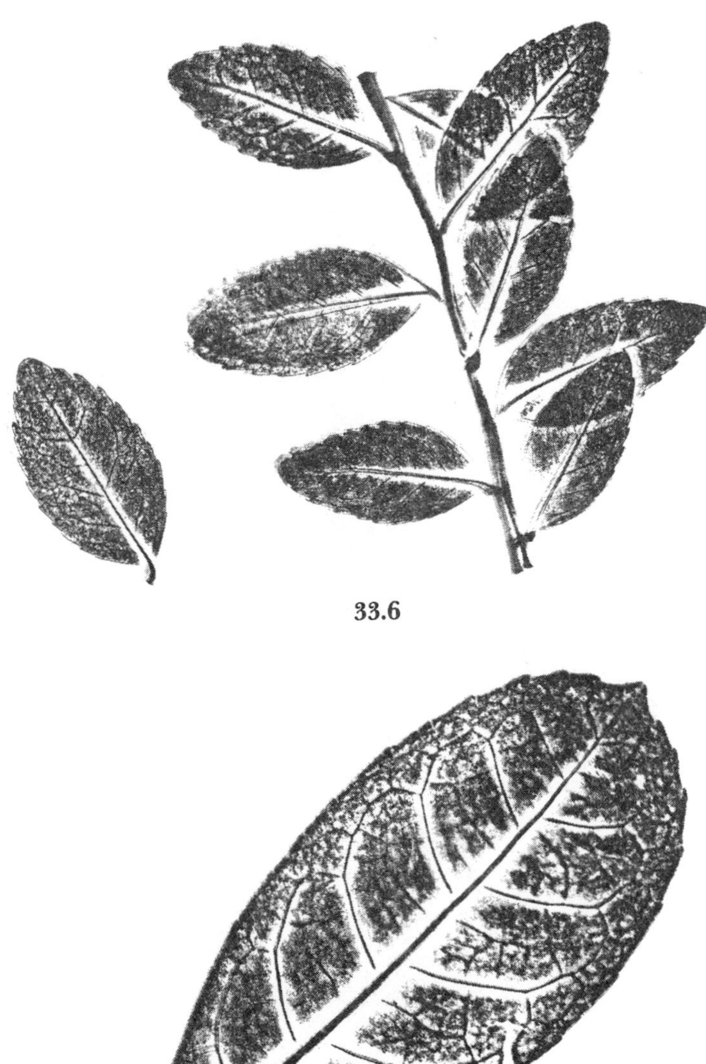

33.6

33.7

This is a family of some 40 genera of trees, shrubs, and climbing vines, whose approximately 400 species are widely distributed over the world. Their fruits are capsules, drupes, and samaras, whose seeds are usually inclosed in pulpy arils.

THE CELASTRUS GENUS (*Celastrus* L.)

The 30 species of this genus are mostly climbers that are useful for covering walls and trellises. They have bright-colored persistent fruits.

34.1 AMERICAN BITTERSWEET (*Celastrus scandens* L.)

Ten species of birds have been recorded feeding on the crimson seeds that are inclosed within the orange-yellow capsules. Unfortunately, many individual plants of this species are found to be sterile and so have little decorative value.

34.1

34.2 ROUND-LEAF BITTERSWEET (*Celastrus orbiculatis* Thunb.)

34.2

THE SPINDLE-TREE GENUS (*Euonymus* L.)

Most of the trees and shrubs in the 120 or so species in this genus possess 4-angled branches and attractive foliage and fruits. The seeds are red, white, or black and are inclosed in an orange-colored aril, as is normal for the family.

34.3 EVERGREEN EUONYMUS (*Euonymus Fortunei* var. *vegetus* Rehd.)

34.4 CORKBARK EUONYMUS (*E. alatus* Sieb.)

A deciduous shrub with broad corky wings on the branches.

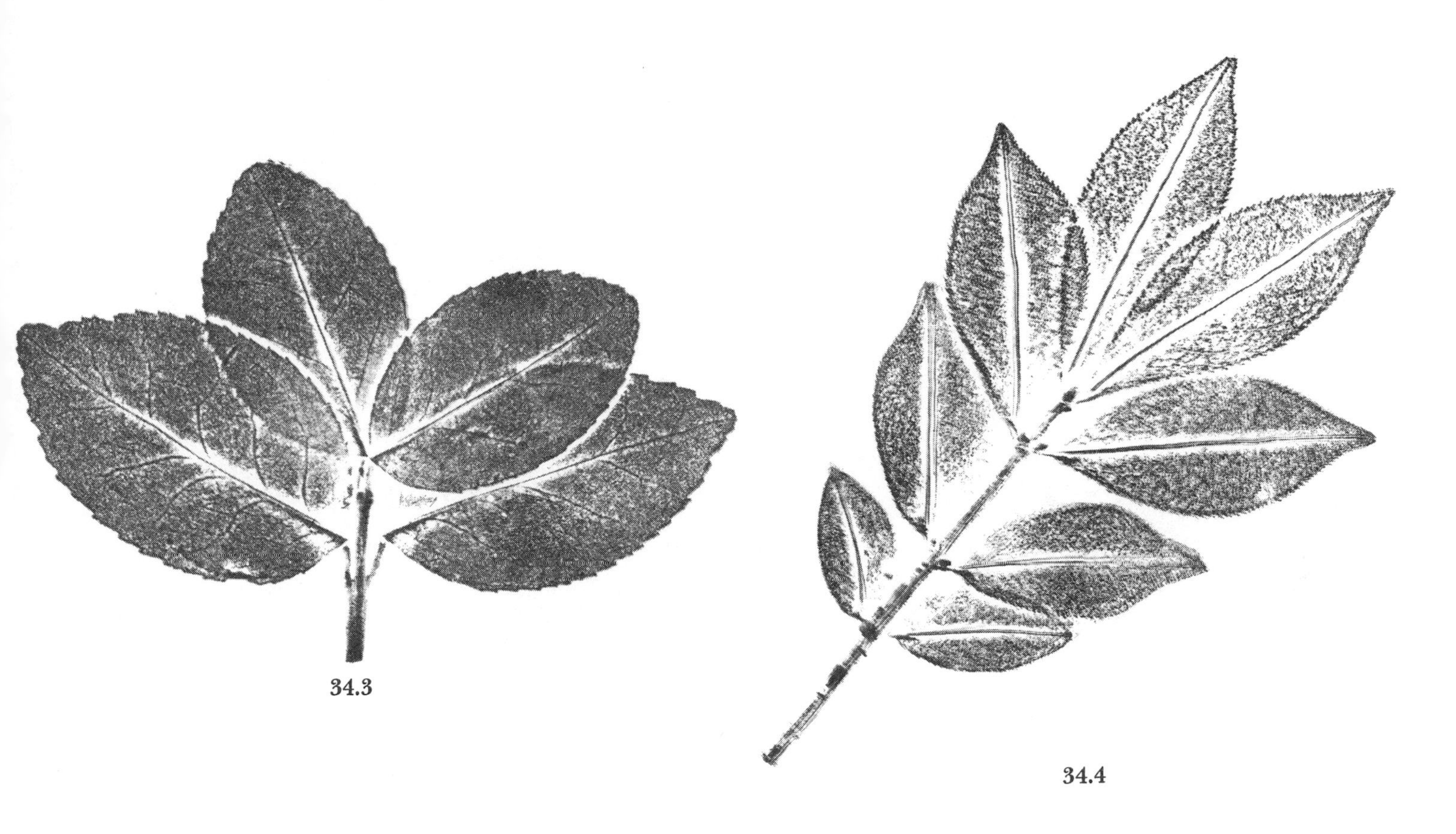

34.3

34.4

34.5 WILD or WAHOO SPINDLE
(*Euonymus atropurpureus* Jacq.)

This species is also called the SPINDLE-TREE and the BURNING-BUSH. (The name "Wahoo" is an Indian term applied to any of several trees and shrubs, especially an ELM.) This tree, being a relative of the BITTERSWEET, has berries that similarly open and become highly decorative. The flowers are purple. The name *Spindle* refers to the tapered rods of wood that are an important part of a spinning wheel.

Johnson states that the bark once had a vogue as a purgative. Hardin and Arena, on the other hand, warn that cases of poisoning have occurred in Europe, resulting in diarrhea, convulsions, and coma. The difference may be only one of dosage, indicating that one man's purgative may be another man's . . . etcetera.

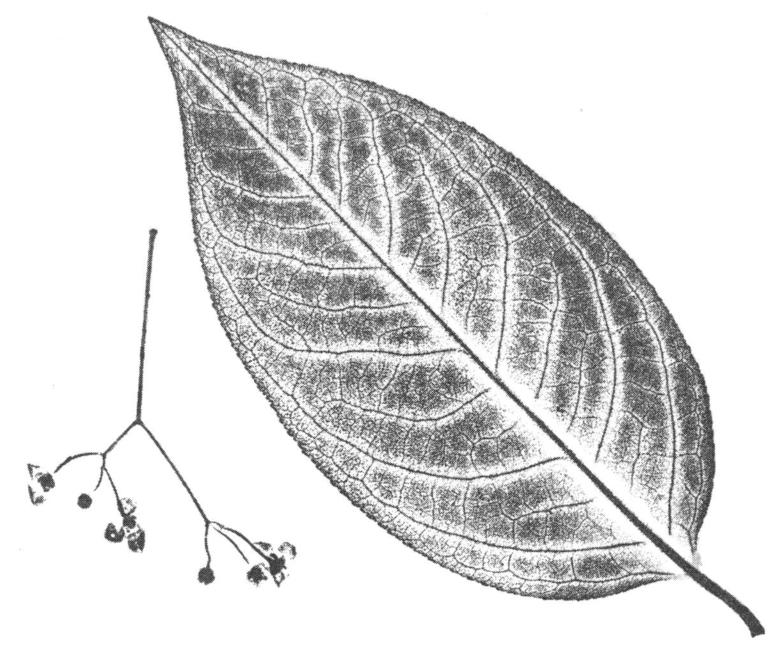

34.5

STAPHYLACEAE

35.1

There are perhaps 5 genera and 25 species of trees and shrubs in this family whose fruits are all leathery or fleshy capsules.

THE BLADDERNUT GENUS
(*Staphylea* L.)

35.1 BLADDERNUT (*Staphylea trifolia* L.)
Ornamental native shrub with bell-shaped white flowers and inflated pods that sometimes turn red. The twigs are streaked in two shades of red. The shiny brown seeds contain a rich-flavored oil, but their shells are so thick that it is impracticable to try to eat them. This plant is not entered in the books as a drug plant, but its properties should be investigated. The live twigs take root and sprout readily in the spring, as this herbalist once discovered when he used some as tomato stakes.

35.1

There are only two genera in the MAPLE family, and most of the over 100 species are to be found in the MAPLE genus. They are trees and shrubs quite widely used for shade and ornament. However, certain species are commercially valuable as sources of lumber for furniture, flooring, interior finish, veneers, and for the manufacture of musical instruments like pianos and violins. Actually, the list of products made from maple wood could be extended considerably. Several species also elaborate a sweetish sap which, when boiled down, produces the maple syrup and sugar of commerce.

The fruit of the Maple is a double samara, the "nose" of childhood.

36.1 SUGAR, HARD, or ROCK MAPLE (*Acer saccharum* Marsh.)

36.2 BLACK MAPLE, BLACK SUGAR MAPLE (*A. nigrum* Michx. f.)

Both of these maples are sources of sugar although *A. saccharum* is the better. In addition, its dried inner bark has been ground and used as breadstuff.

36.2

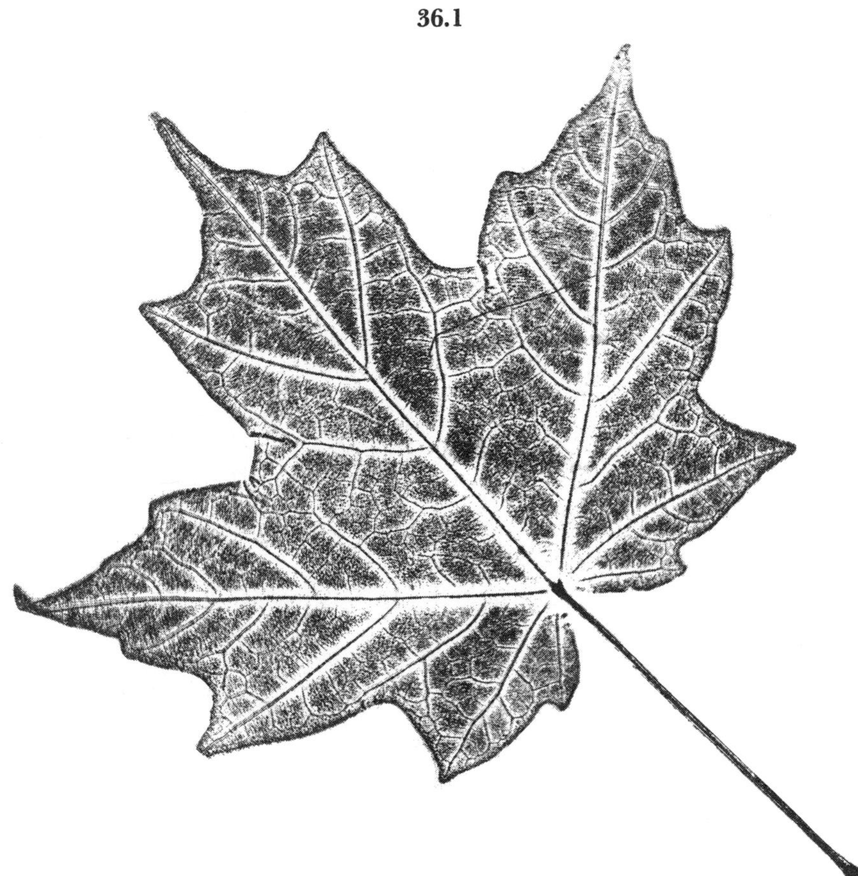

36.1

36.3 NORWAY MAPLE (*A. platanoides* L.)
If there should be any confusion of this leaf with that of the SUGAR MAPLE, one need only break the petioles of each to make a distinction: The NORWAY MAPLE exudes a milky juice; the SUGAR MAPLE does not.

36.4 SILVER MAPLE (*Acer saccharinum* L.)

36.5 SWAMP-RED MAPLE (*A. rubrum* L.)

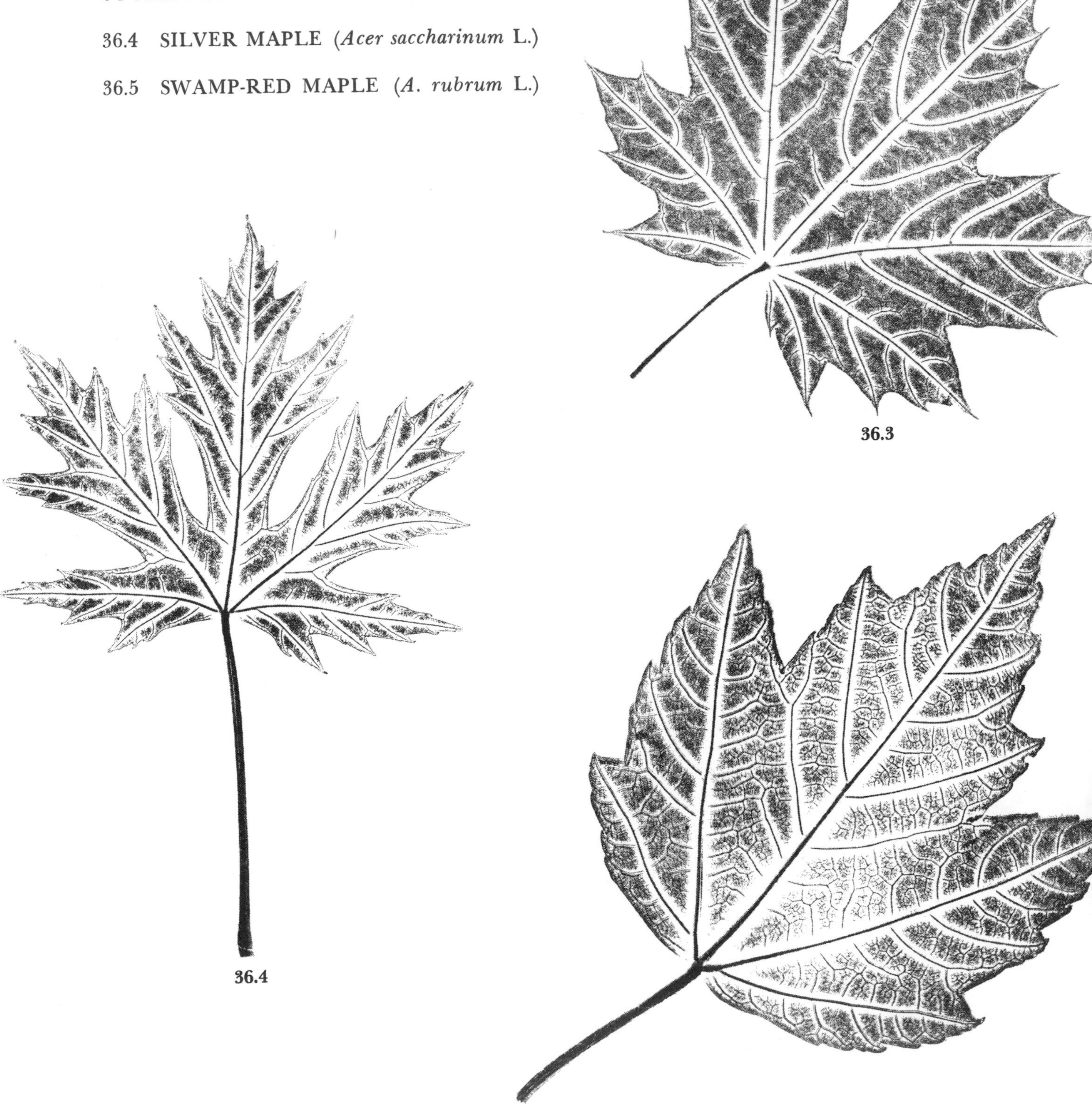

36.3

36.4

36.5

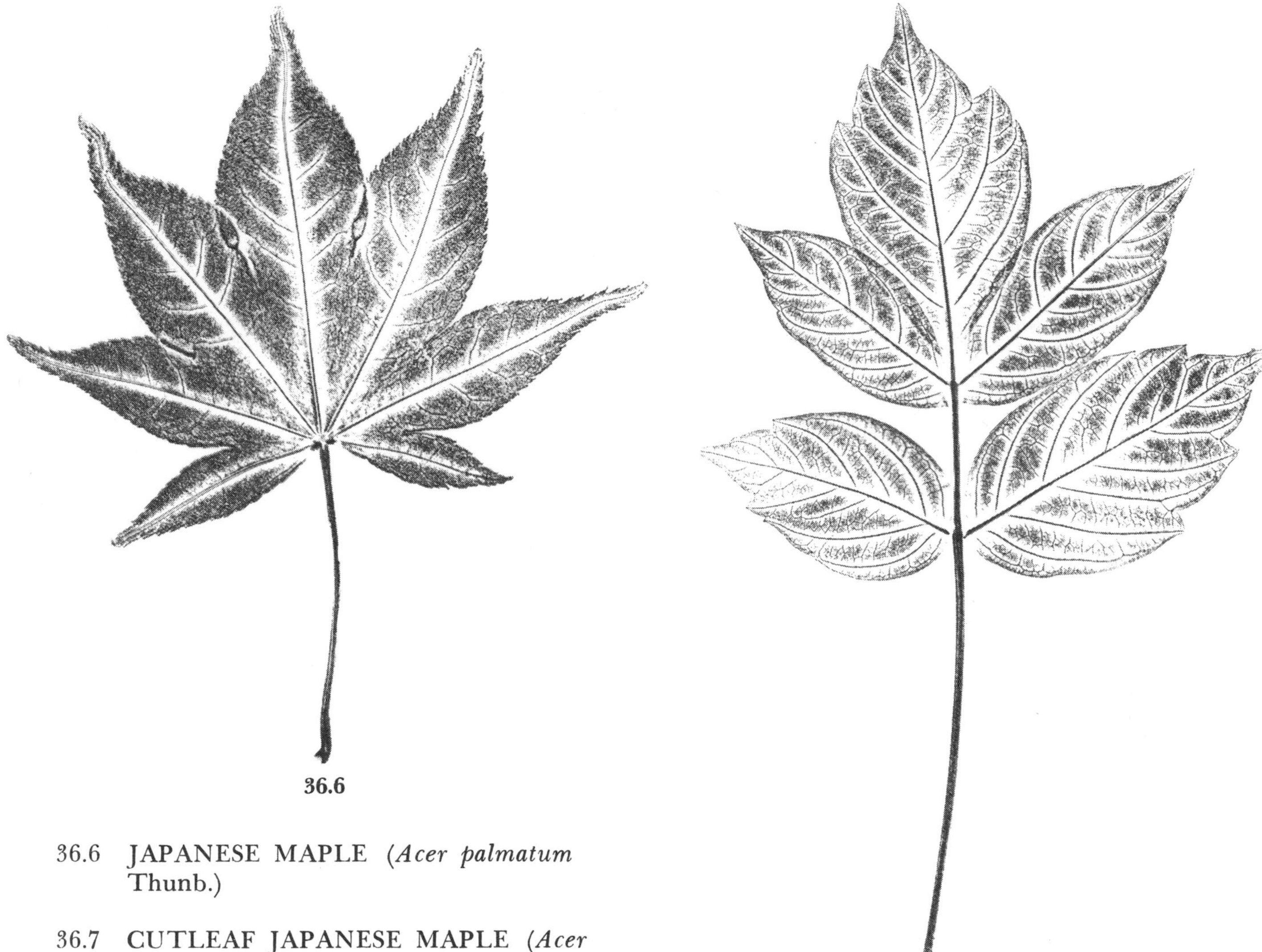

36.6

36.8

36.6 JAPANESE MAPLE (*Acer palmatum* Thunb.)

36.7 CUTLEAF JAPANESE MAPLE (*Acer palmatum* var. *dissectum* Maxim.)

36.8 ASHLEAF MAPLE; BOX-ELDER (*Acer Negundo* L.)

36.9 STRIPED MAPLE: MOOSEWOOD (*Acer pensylvanicum* L.)

36.7

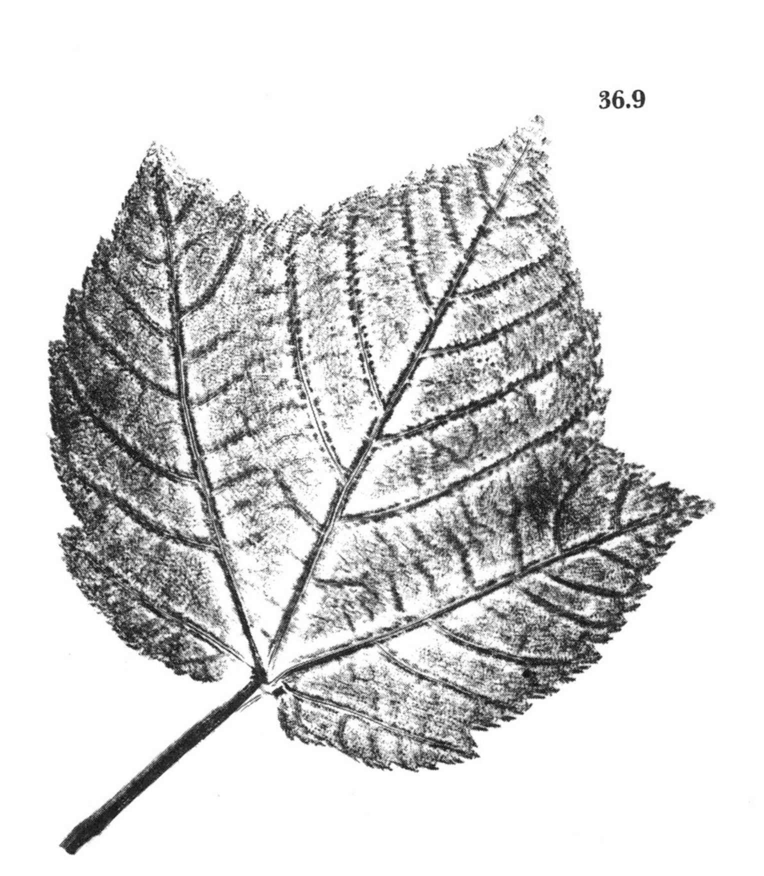

36.9

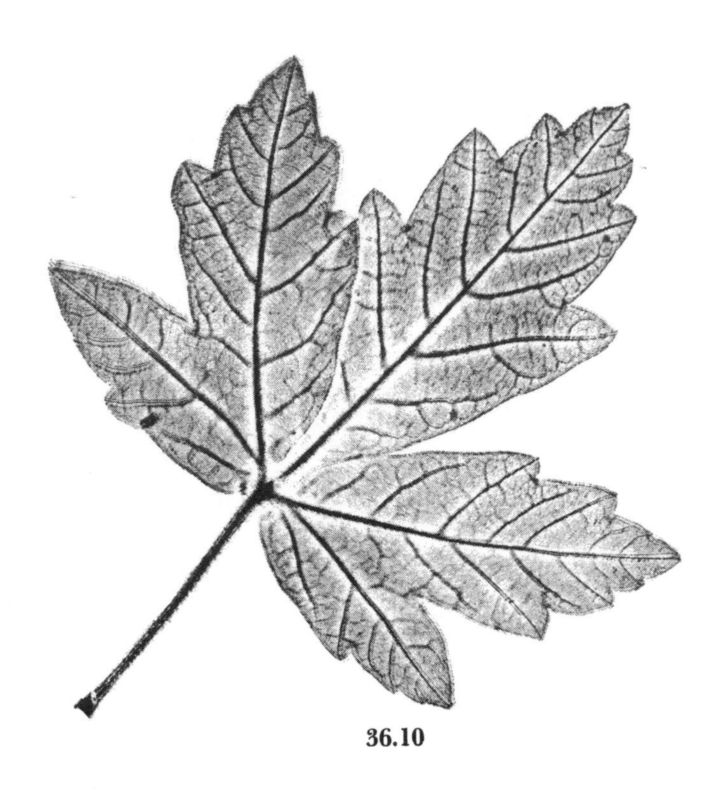

36.10

36.12

36.10 PAPERBACK MAPLE (*Acer griseum* Pax)

36.11 SYCAMORE MAPLE (*Acer Pseudo-Platanus* L.)

This is *the* Sycamore of English literature and tree books. They refer to our Sycamore (*Platanus occidentalis*) as the American Plane-tree.

36.12 AMUR MAPLE (*Acer Ginnala* Maxim.)

36.11

36.13a and 36.13b HEDGE MAPLE (*Acer campestre* L.)

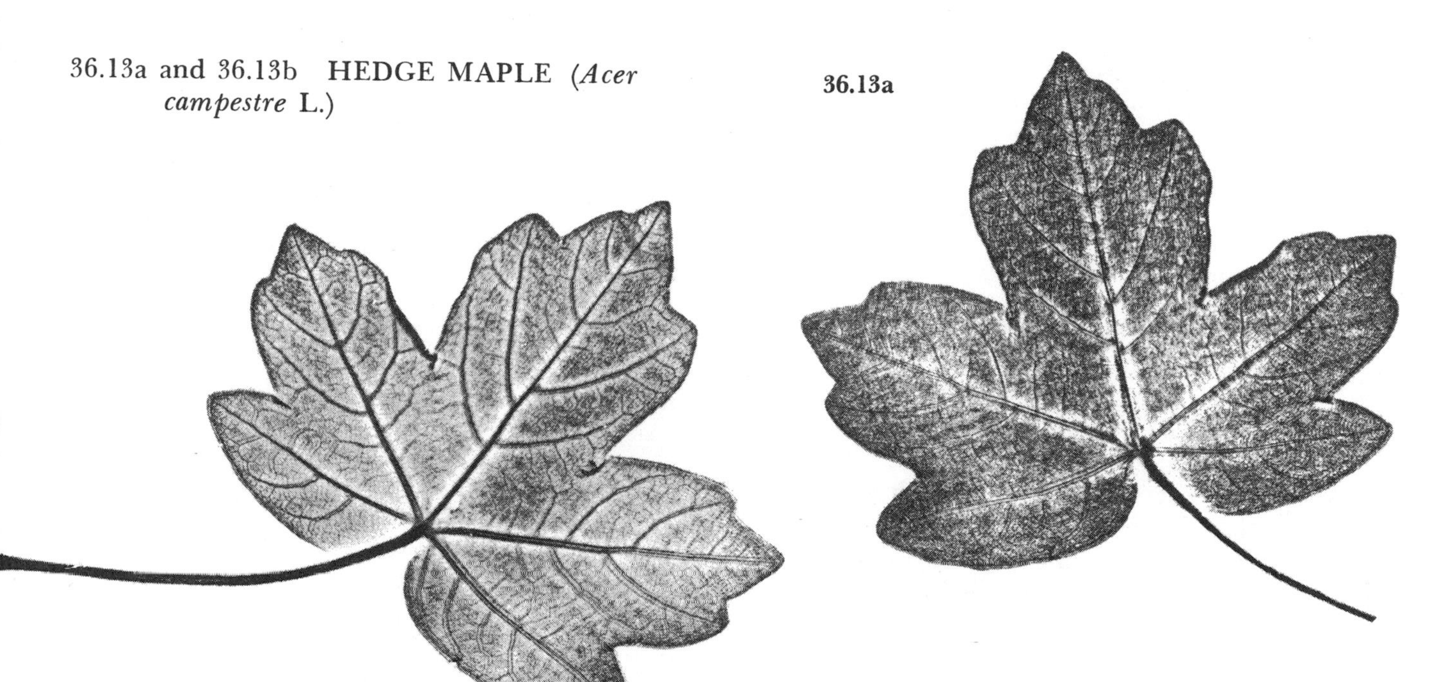

36.13a

36.13b

HIPPOCASTANACEAE

This family of trees and shrubs with palmately compound leaves occurs widely throughout the north temperate zone. It consists of 3 genera and over 25 species. The fruits are large leathery capsules containing from 1 to 3 large shining seeds.

THE HORSE-CHESTNUT (BUCKEYE) GENUS (*Aesculus* L.)

About 25 species of shade trees with showy flowers. Of the species shown, both *Ae. Hippocastanum* and *Ae. carnea* have resinous winter buds, but the remaining three species do not.

While the seeds of this genus are toxic when eaten raw, they can be prepared after the manner suggested for acorns (roasted, mashed, leached, and dried) to produce an edible meal. Eating the seeds raw, however, as well as drinking a tea made from the leaves or twigs may result in a variety of symptoms, culminating in paralysis and stupor. In Europe an alcoholic extract has been used as a treatment for hemorrhoids, and in bygone days in the U.S. the seeds were hung around the neck on a string to prevent rheumatism.

37.1 COMMON or WHITE HORSE-CHESTNUT (*Aesculus Hippocastanum* L.)

The flowers are white, the pods green, the buds sticky. A red-brown dye can be extracted from the husks.

37.2 PINK HORSE-CHESTNUT (*Ae. carnea* Hayne)

The flowers are pink to red; the leaves are wrinkly, not flat.

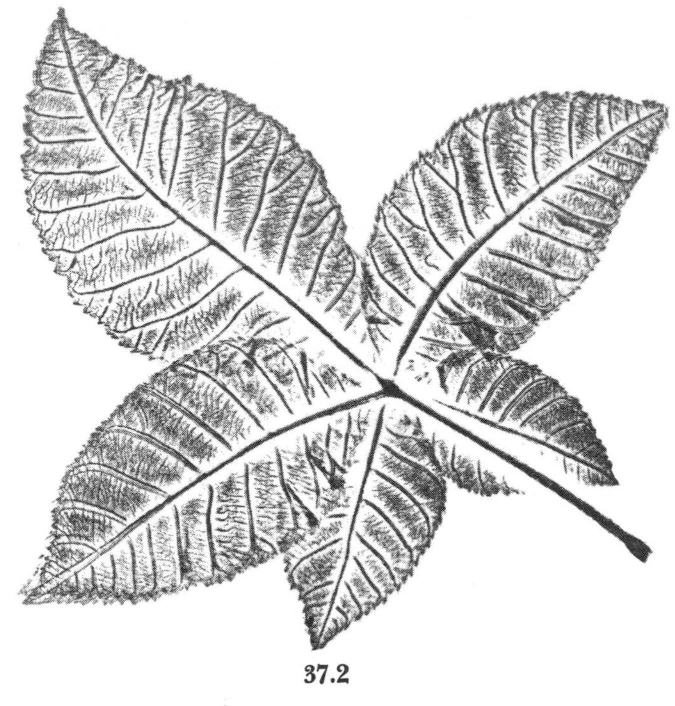

37.2

37.1

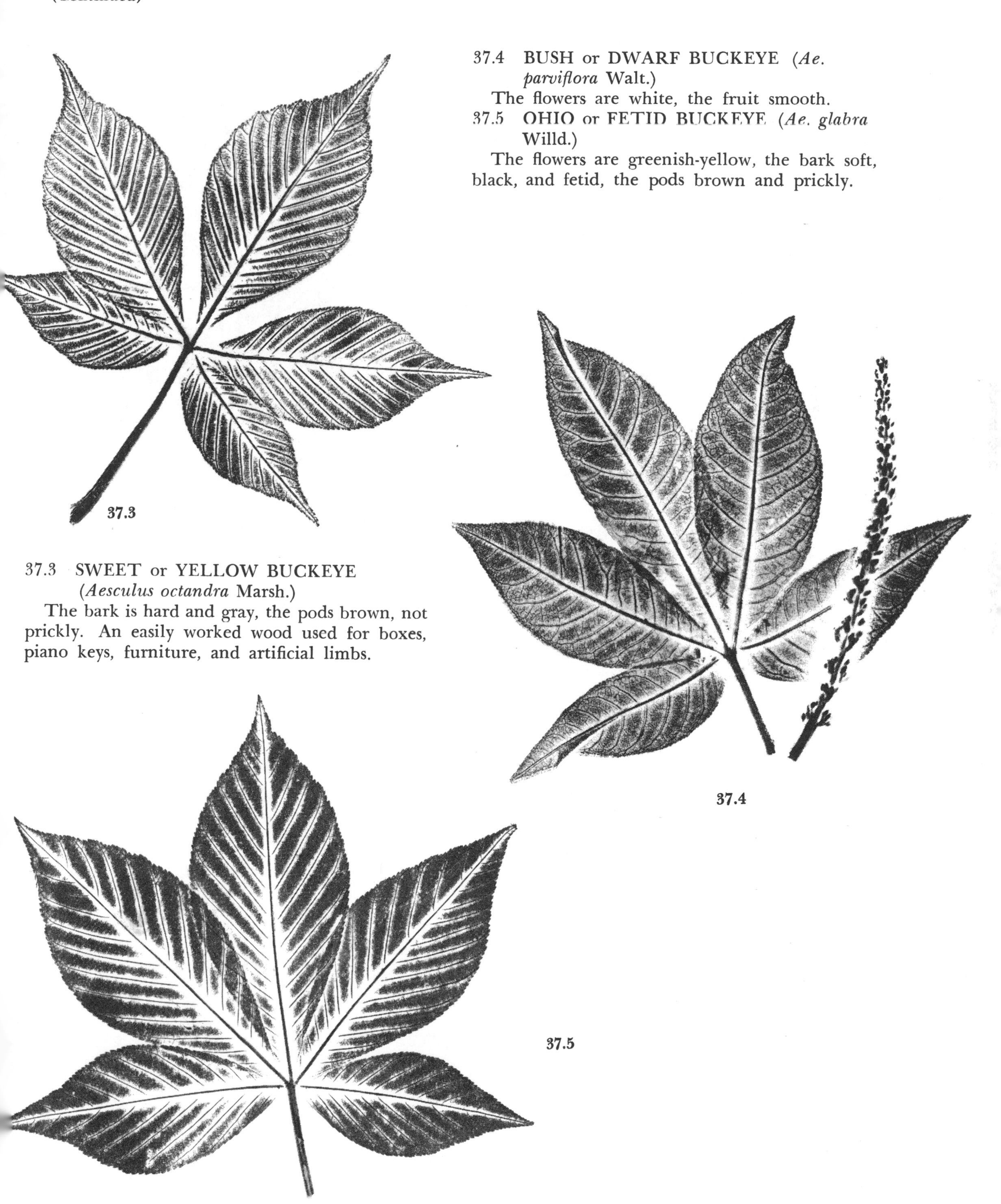

37.3 SWEET or YELLOW BUCKEYE (*Aesculus octandra* Marsh.)

The bark is hard and gray, the pods brown, not prickly. An easily worked wood used for boxes, piano keys, furniture, and artificial limbs.

37.4 BUSH or DWARF BUCKEYE (*Ae. parviflora* Walt.)

The flowers are white, the fruit smooth.

37.5 OHIO or FETID BUCKEYE (*Ae. glabra* Willd.)

The flowers are greenish-yellow, the bark soft, black, and fetid, the pods brown and prickly.

In the 125 genera and approximately 1,000 species of this subtropical family are many ornamentals and a few fruit trees of minor importance. The SOAPBERRY genus (*Sapindus*) counts among its 15 or so species one whose brown shining berries produce a soap-like lather when agitated in water. This is the FLORIDA SOAPBERRY (*Sapindus Saponaria*).

THE KOELREUTERIA GENUS
(*Koelreuteria* Laxm.)

Introduced into the U.S. from China, Japan, and Korea. One of the four species in the genus has become well known as an ornamental tree. It has compound leaves and round black seeds contained in a papery capsule that opens into three valves.

38.1 HARMONY-TREE; GOLDENRAIN-TREE (*Koelreuteria paniculata* Laxm.)

This is one of the few yellow-flowering trees. It is tolerant of a wide range of soils and survives the heat and fumes of cities.

38.1

THE MELICOCCA GENUS (*Melicocca* L.)

There are but two species in this genus and of them only one is grown for its fruit in the warmer portion of the U.S. This fruit is a drupe with fleshy pulp that incloses a seed surrounded by a large fleshy aril.

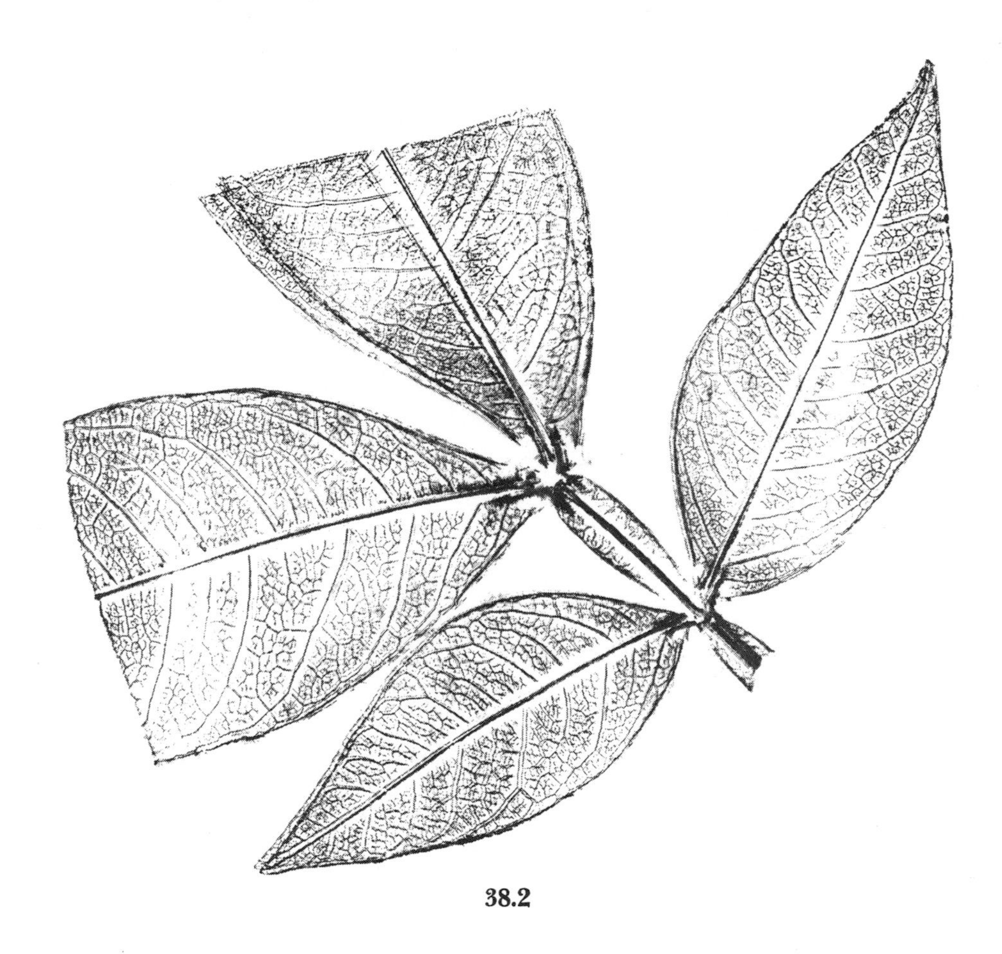

38.2

38.2 SPANISH-LIME (*Melicocca bijuga* L.)
The large round seed of this tree is often roasted like a chestnut.

LITCHI; LEECHEE, etc. (*Litchi sinensis* Sonn.)
This is another fairly familiar member of the Soapberry family. The nuts have often been for sale in Chinese restaurants and laundries.

The family contains both endemic American species and some that were imported from the Old World. There are about 50 species and over 600 genera of erect or climbing and often spiny trees, shrubs, and a few herbaceous plants in the group. Some produce edible berries and most are useful in conservation projects.

THE JUJUBE GENUS (*Ziziphus* Mill.)

This unusually named genus contains some 40 species of deciduous and evergreen trees and shrubs. Some are ornamental; others have edible fruits.

39.1 JUJUBE-TREE (*Ziziphus Jujuba* Mill.)

The fruit of the JUJUBE-TREE is a dark red or brown drupe about one inch long. Its whitish flesh encloses a hard, 2-celled stone. The fruits can be eaten raw, dried, candied, or in jam.

39.1

THE BUCKTHORN GENUS (*Rhamnus* L.)

Of the 600 species in the family, 100 are in this genus. In some genera the berries produced are poisonous or medicinal and in others edible. The leaves of these shrubs can be recognized by their few, strongly arched veins.

39.2 COMMON BUCKTHORN (*Rhamnus cathartica* L.)

Usually Buckthorn shrubs or small trees are found scattered through the woods, but it is also used for hedging and is sometimes seen as a specimen tree of very symmetrical form. The berries are medicinal and should be considered poisonous, never eaten. They can be used to produce a dye.

39.3 FRANGULA BUCKTHORN: ALDER BUCKTHORN (*R. Frangula* L.)

The inner bark of all BUCKTHORNS is used as a cathartic, but see below.

39.2

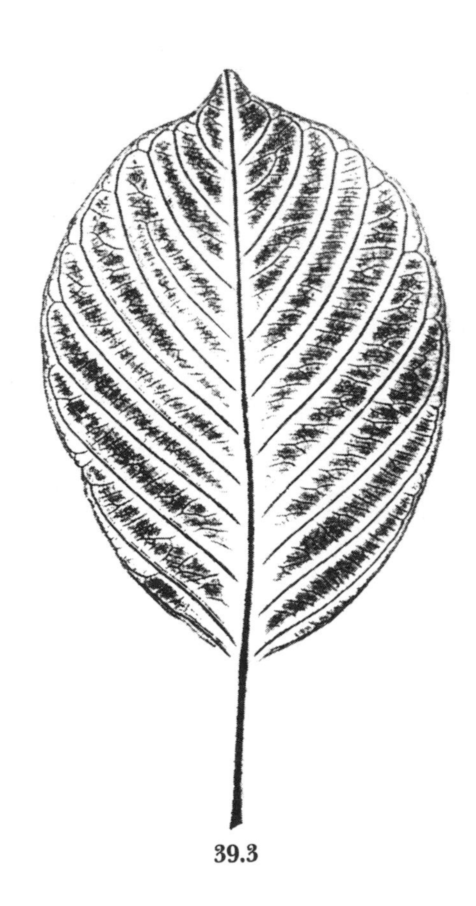

39.3

CASCARA SAGRADA (*R. Purshiana* DC.)

The most widely used of the BUCKTHORNS for medicinal purposes. The reddish-brown "Sacred Bark" of the Indians and pioneer Spanish settlers is still being used. During the summer long strips of bark are peeled from the trees and stored for a year before being made into a laxative and tonic.

The 35 genera and 370 or more species of trees, shrubs, and a few herbs in this family are characterized by a fibrous and mucilaginous bark.

THE LINDEN, BASSWOOD, OR LIME GENUS (*Tilia* L.)

In the north temperate zone there are about 30 native species belonging to the LINDEN genus. They are famous as shade trees, perhaps none more famous than those indicated by the Berlin street-name *Unter den Linden.*

The wood of the Lindens is easy to work and yet is very strong. It is used, for instance, for the frames within which bees place the honeycomb in the hive, because it can successfully withstand the required right-angle bending. Furthermore, the very fragrant flowers of these trees are rich in nectar, so that the production of honey may well both begin and end in association with the Linden tree. The humming of bees gathering nectar may be so loud that the tree can be "heard" some distance away. The wood is also useful in campcraft programs since it can be carved with a penknife almost without regard to the direction of the grain. The inner bark of the LINDEN is the source of very strong fibers, from which I have many times made great lengths of rope. The bark is stripped from a tree that hopefully has been felled for some other sufficient reason. It is gathered into bundles and tied together. Rocks are then used to submerge the bundles in a lake or stream. In due time, depending largely on the temperature of the water, the inner fibers begin to separate from the corky and woody outer bark. These fibers are removed from the nonfibrous parts by hand. They are further separated from each other by running the fingers through them like a comb. When dried they can be plaited, or better, twisted, into a two-stranded rope of any diameter or length. The length is achieved by the gradual feeding in of more fibers as the rope-maker reaches the end of the first ones.

40.1 AMERICAN LINDEN (*Tilia americana,* L.)

40.1

40.2 EUROPEAN LINDEN (*Tilia europaea* L.)

40.3 WHITE LINDEN (*T. tomentosa* Moench.)
Not a native tree, but often seen planted in parks. The leaves are dazzling green above, white beneath.

40.3

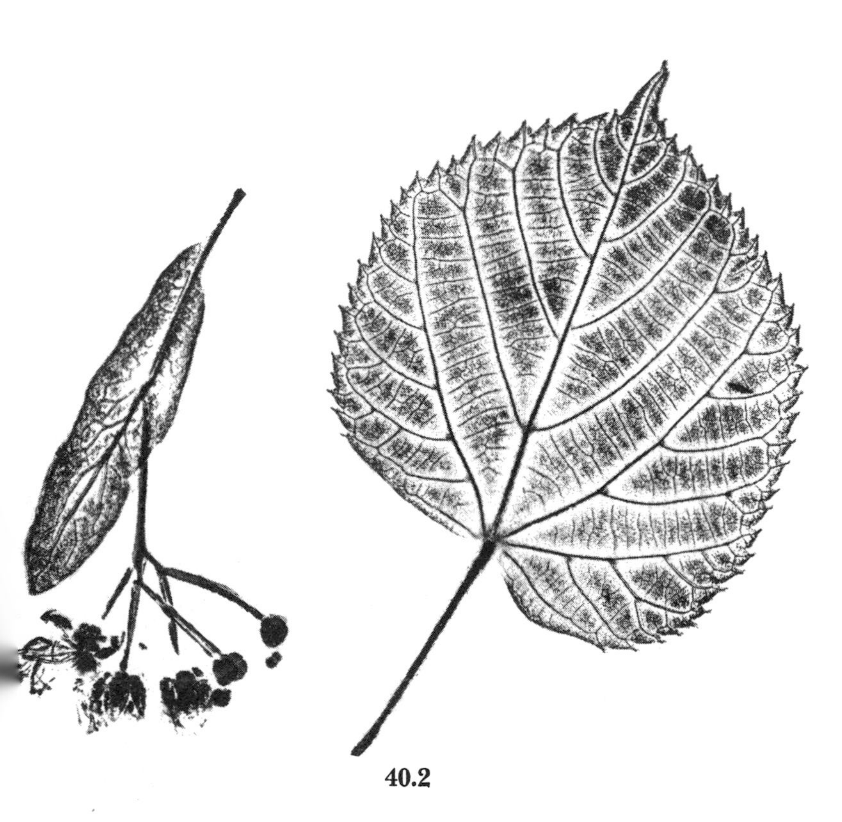

40.2

About 1,000 species are included in the 40 to 50 genera of herbs, trees, and shrubs in the Mallow family. Some are ornamental, a few are food producing, but perhaps the most famous are those that produce the cotton fibers of commerce. The leaves in the family are palmately veined.

THE COTTON GENUS (*Gossypium* L.)

There are more than 20 species and very many hundreds of cultivated varieties of plants that are characterized by the elaboration of hair-cells on their globular seeds. *G. hirsutum,* which is shown here, is an "upland" cotton in the form of a shrubby, much-branched annual. The flowers are light yellow, or white and without spots. The bolls dehisce into 4 or 5 valves. The hair-cells on the seeds within are the basis of the cotton textile industry.

41.1 UPLAND COTTON (*Gossypium hirsutum* L.)

41.1

THE ROSE-MALLOW GENUS (*Hibiscus* L.)

The description of this genus parallels very closely that of the family's description above. There are ornamental, food- and fiber-yielding herbs, trees, and shrubs in the 200 or so species included in it.

41.2 ROSE-OF-SHARON; ALTHEA (*Hibiscus syriacus* L.)

41.3 ROSE-OF-CHINA; (FLORIDA) HIBISCUS (*H. Rosa-sinensis* L.)

41.2

41.3

Some vines, as well as trees and shrubs, are found in this family of about 50 genera. And among its 750 or so species are several that provide products dearly beloved by children and adults alike, namely, cocoa and chocolate from trees in the genus THEOBROMA and cola from the genus COLA.

THE THEOBROMA GENUS
(*Theobroma* L.)

This genus has been named with the Greek equivalent of "Food of the Gods." It consists of about 20 species of small trees that are widely grown in tropical America. The fruits of these trees are large woody drupes or pods. In *Theobroma Cacao,* an evergreen tree, the pods are one foot long and contain 1-inch-wide flat seeds. These are the seeds from which cocoa and chocolate are manufactured. The flowers, and the pods that follow, are borne on the main trunk of the tree as well as on the branches. The ripe pods are yellow or red and contain a whitish mucilage enclosing a mass of white seeds. The mucilage and the beans are fermented on the ground in piles that are covered by banana leaves or in specially designed wooden boxes. When the fermentation process is complete, the beans are dried either in the sun or in mechanical dryers. They are then shipped to the importing countries where the cocoa and chocolate are produced.

42.1 CHOCOLATE-TREE; CACAO
(*Theobroma Cacao* L.)

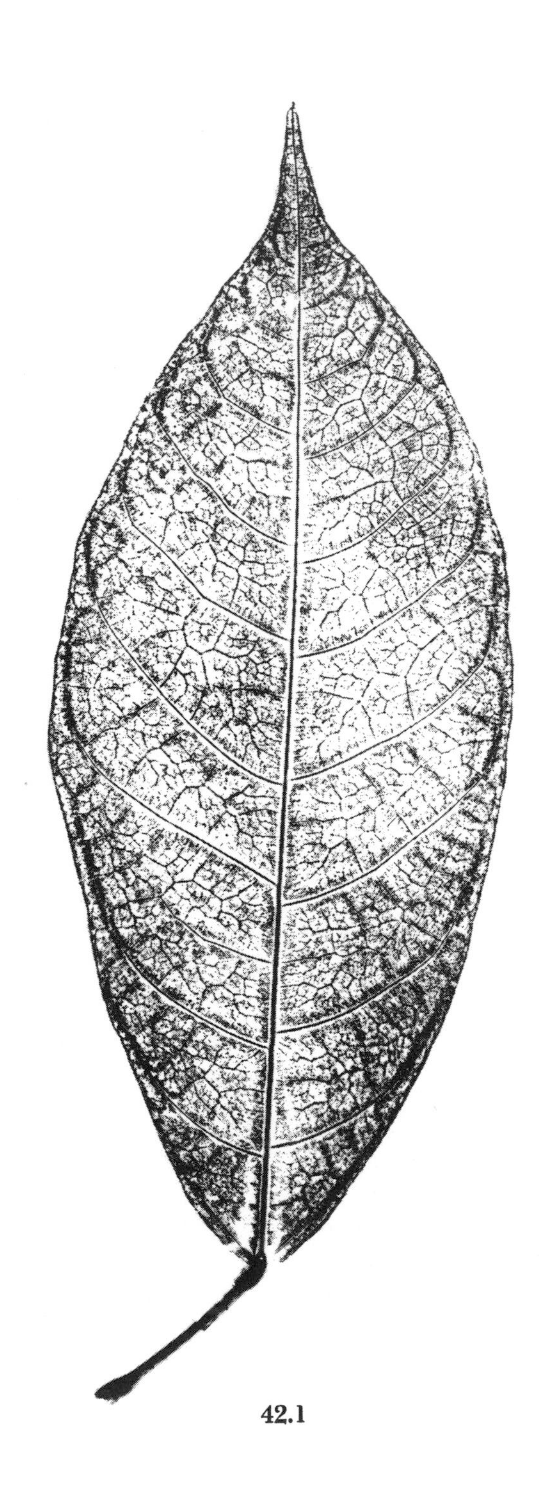

42.1

Of the 18 genera and about 200 species in this family, some are used for ornamental planting and one genus provides the tea of commerce. The leaves of all are rather leathery.

THE TEA GENUS (*Thea* L.)

The 16 species classified in this genus are evergreen trees and shrubs. The species shown below, *Thea sinensis,* grows wild from India to China. It can be cultivated on plantations, however, from the equator to the Black Sea as well as in some portions of northern China and most of Japan. It prefers an acid soil and requires considerable rainfall. At intervals of a week or so a tea picker removes the terminal bud and the next two (or for lower-quality tea, three) leaves from the end of each shoot. The next steps in the processing of the leaves must occur rather quickly and they include withering, rolling, fermenting, drying, and grading. Tea leaves contain both caffeine and tannin.

43.1 TEA (*Thea sinensis* L.)

43.1

THE STEWARTIA GENUS (*Stewartia* L.)

Some of the 8 species in this genus are native in eastern North America. They are noted for their showy flowers which, in the species shown below, are 3 to 4 inches in diameter. The purple filaments and bluish anthers of the flowers contrast strongly with their white petals.

43.2 STEWARTIA (*Stewartia Malachodendron,* L.)

43.2

THE FRANKLINIA GENUS
(*Franklinia* Marsh.)

There is only one species in this entire genus that was named in honor of Benjamin Franklin. The tree has not been found wild since 1790, but is seen occasionally in cultivation. Its white flowers are some 3 inches across. The pattern of its branching is very interesting since the terminal 8 to 10 inches of the branchlets arch strongly upward.

43.3 FRANKLINIA (*Franklinia alatamaha* Marsh.)

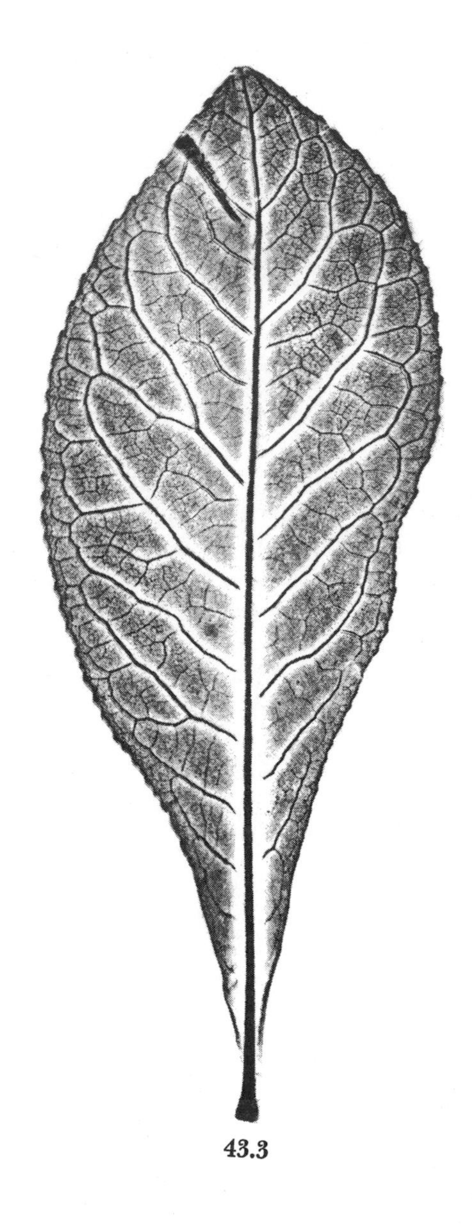

43.3

The TAMARISKS are small trees and shrubs in four genera and about 100 species. Some are planted here as ornamentals.

THE TAMARISK GENUS (*Tamarix* L.)

Consists of some 75 species, many of which are so similar in appearance that only technical characteristics can be used to separate them. Some species are planted as windbreaks in the deserts of North Africa.

44.1

44.1 FRENCH TAMARISK (*Tamarix gallica* L.)

Although native in western Europe, it has now escaped in Texas, Arizona, and New Mexico.

There are only two genera of small trees in this family, which consist of some 30 species. One of the two genera, CARICA, is noted for its edible fruit. In both the wood is soft and the sap milky. Their trunks, being unbranched, have a palm-like appearance.

THE PAWPAW GENUS (*Carica* L.)

In this genus there are about 25 species, some of which may be seen growing as ornamentals in greenhouses but also are frequently found under cultivation in the tropics. They are farmed for their edible fruits and for the papain which is extracted from these fruits. David S. Marx was among the first to realize that a meat tenderizer could be produced from papain. He suggested the idea to his brother, who produced such a preparation commercially for some time.

45.1 PAPAYA; PAWPAW (*Carica Papaya* L.)

The leaves of this tree are sometimes two feet across and the fruit may vary from 3 to 20 inches in length. The fruits are large berries with thick yellow flesh and a number of blackish seeds.

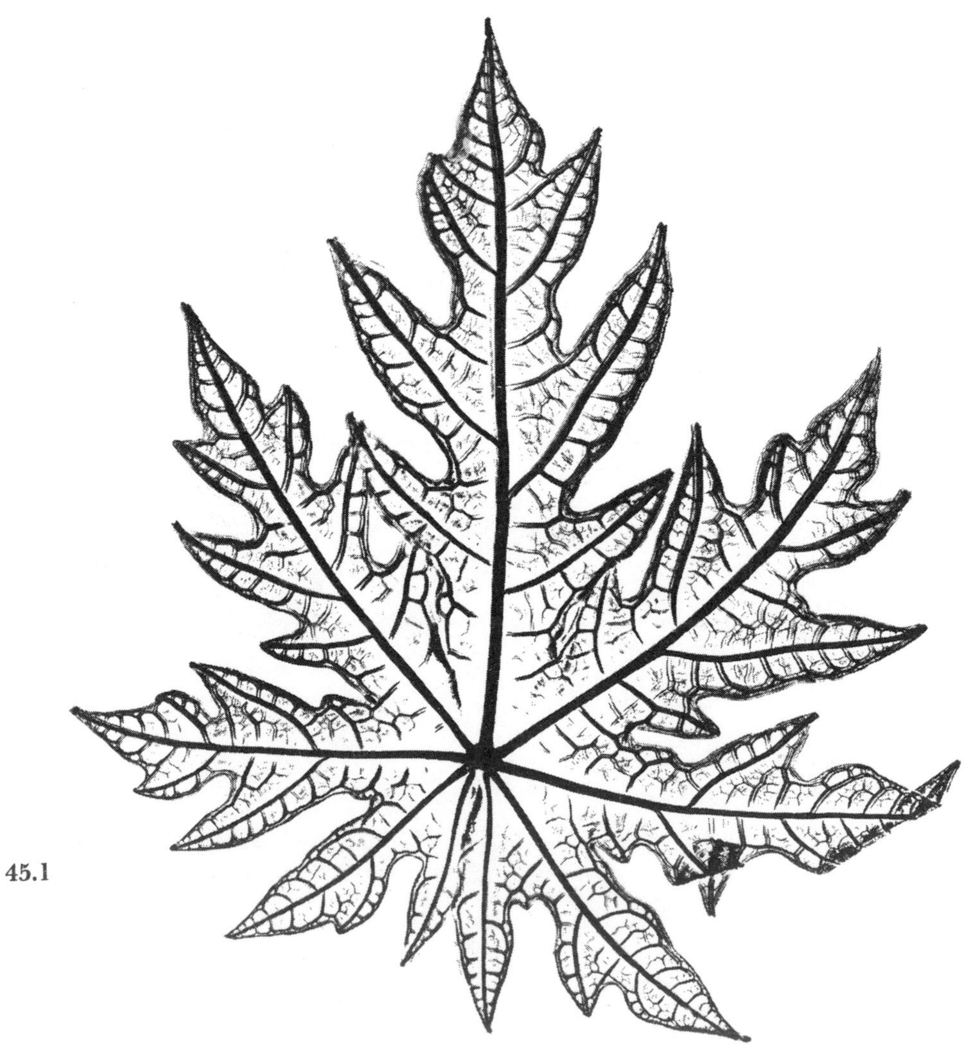

45.1

In the temperate and subtropical regions of the northern hemisphere we find 3 genera with more than 45 species of OLEASTERS. These trees and shrubs are largely grown as ornamentals but a few bear edible fruits. These fruits are achenes or nuts enclosed by a fleshy receptacle.

46.1

THE ELAEAGNUS GENUS (*Elaeagnus* L.)

This is a genus of evergreen and deciduous small trees and shrubs, some of which are spiny, whose branchlets are thickly covered with silvery or brownish scales. There are about 40 such species in the genus.

46.1 OLEASTER (TREBIZONDE-DATE; RUSSIAN OLIVE) (*Elaeagnus angustifolia* L.)

In this species the branchlets are silvery white and the ½-inch-long fruits, which are covered with silvery scales, have a yellow flesh.

This deciduous small tree or shrub has such unique characteristics that it has been placed in a family of its own. There is, therefore, just the one genus (*Punica*), which, however, is divided into two species. The fruit of the Pomegranate is a hard, leathery-skinned berry internally divided into several cells. Within the compartments are numerous seeds, each enclosed in a reddish to pink acid-sweet pulp. Eating a pomegranate has its difficulties, since there is so little reward (the pulp) for the effort involved. Some recipes call for only the juice to be used in cooling drinks or to be made into wine.

The tree is quite hardy and so may be raised as an ornamental for its flowers, but in order to ripen the fruit a warm climate is required.

Dwarf varieties are sold by some nurseries (*P. Granatum* var. *nana*).

47.1 POMEGRANATE (*Punica Granatum* L.)

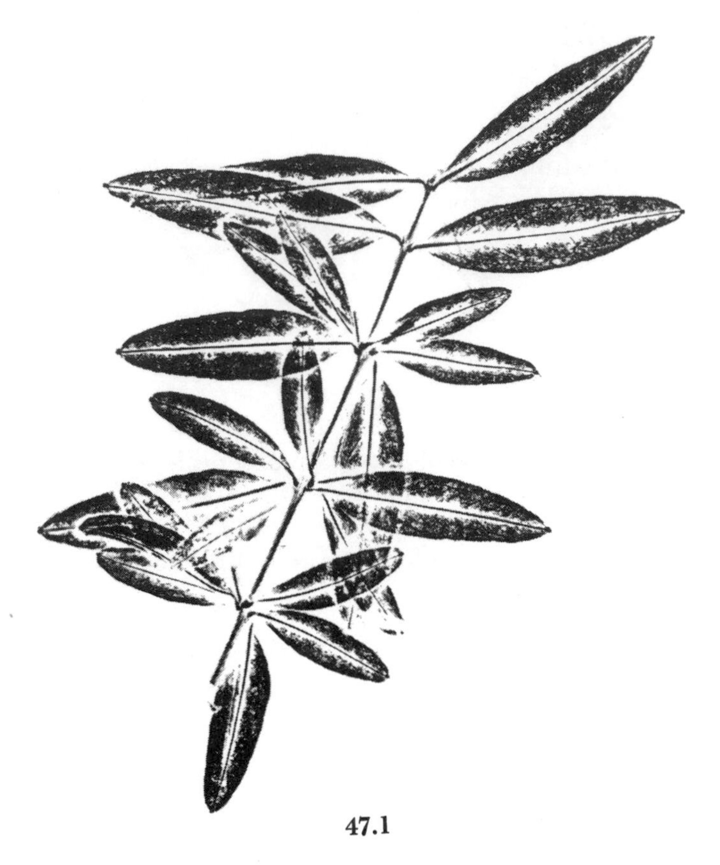

47.1

There are 3 genera and 8 species of deciduous trees native in North America and Asia in this family mostly planted as ornamentals, although the wood of the TUPELOS is used in veneers, flooring, and in other instances where toughness and resistance to wear are important.

THE TUPELO GENUS (*Nyssa* L.)

A genus of 6 species, only two of which are at all common in the U.S.

48.1 TUPELO-TREE; PEPPERIDGE; SOUR or BLACK GUM (*Nyssa sylvatica* Marsh.)

A tree of swampy habitat with beautiful autumnal foliage; the fruit is a dark, blue drupe. Its wood is very difficult to split.

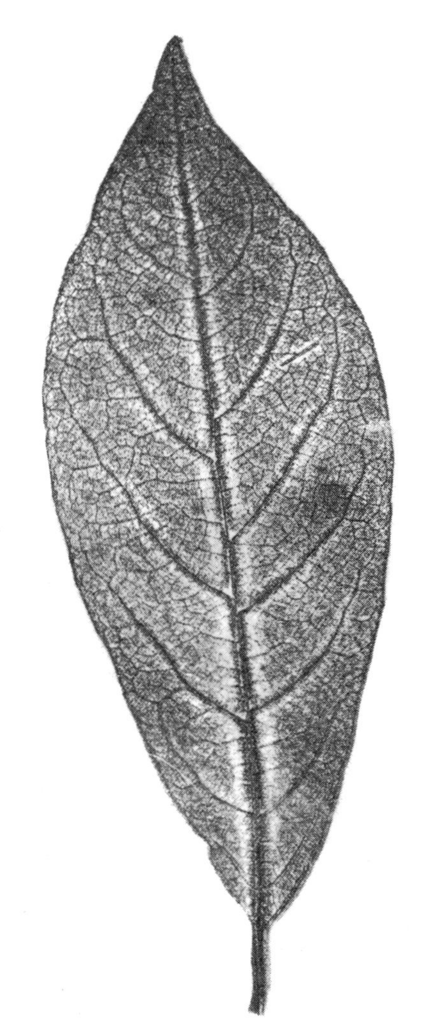

48.1

THE DAVIDIA GENUS (*Davidia* Baill.)

A single-species genus whose tall (60 ft.) trees have 3- to 6-inch creamy white flowers.

48.2 DOVE-TREE (*Davidia involucrata* Baill.)

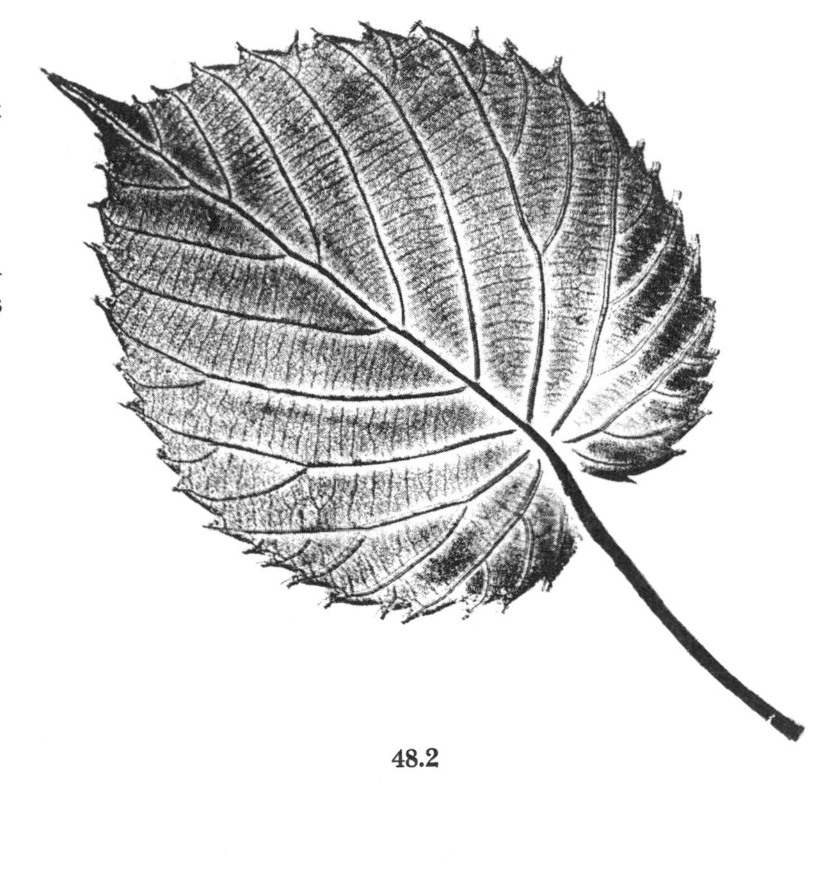

48.2

A rather diverse family of about 75 genera and 3,000 species, but all with aromatic fragrance. Many are grown for their ornamental use in parks and other public displays; some are grown for their edible fruits. The fruits are of several kinds, including berries, drupes, capsules, and nuts.

THE BOTTLE-BRUSH GENUS
(*Callistemon* R.Br.)

Although the 25 species of this genus are native in Australia, they are often seen as ornamentals in the warmer portions of the U.S.

49.1 BOTTLE-BRUSH TREE (*Calistemon rigida* R.Br.)

The leaves are but 1/8th-inch wide, rigid and sharp-pointed. The mid- and marginal veins are prominent.

THE EUCALYPTUS GENUS
(*Eucalyptus* L'Her.)

In Australia and Malaysia there are many valuable timber trees included among the 300 species that are members of this genus. At the present time many of the species are offered for sale in California.

49.2 TASMANIAN BLUE GUM (*Eucalyptus Globulus* Labill.)

Up to 300 feet tall. The shedding of the bark leaves the trunk a bluish-white.

49.1

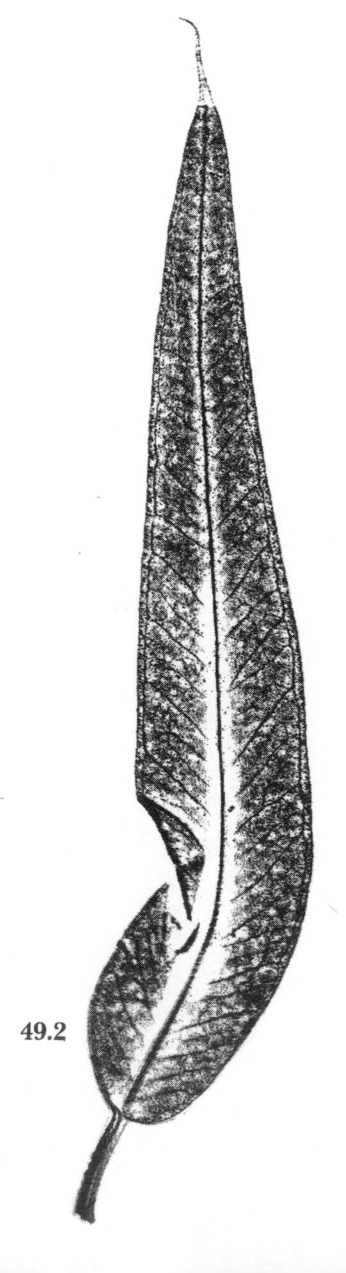

49.2

THE PSIDIUM GENUS (*Psidium* L.)

About 150 species of trees and shrubs that are cultivated for their edible fruits.

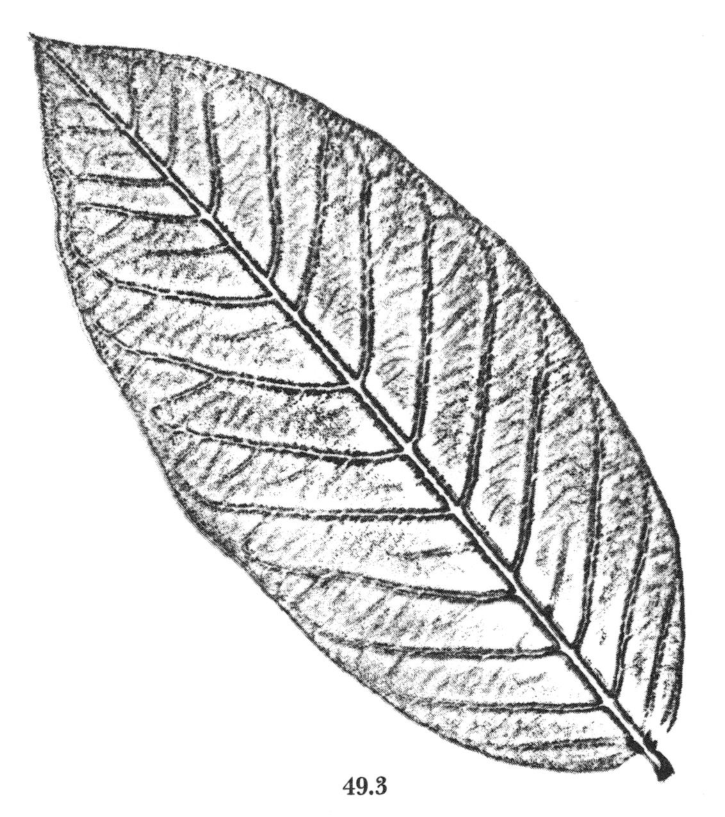

49.3

49.3 GUAVA (*Psidium Guajava* L.)

These small trees have spread out from their original habitat and are now one of our most commonly planted tropical fruits. Guavas are higher in vitamin C content than citrus fruits. While some people enjoy eating them raw, their best use is stewed or made into jams and jellies.

The smooth reddish-brown bark peels off the younger branches in a characteristic way. The smaller branchlets are 4-angled.

49.4 STRAWBERRY GUAVA (*P. Cattleianum* Sabine)

The fruits of this tree are of a reddish-purple color, smaller than those above and of good flavor. The branchlets are circular in cross-section.

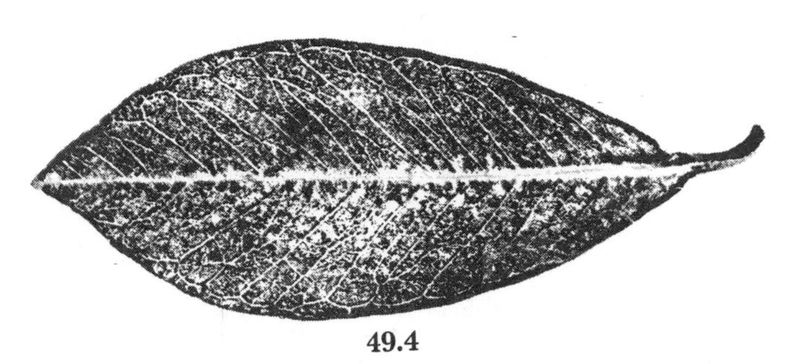

49.4

THE BOTTLE-BRUSH GENUS (*Melaleuca* L.)

About 100 species of Australian ornamental and timber trees. Some species are useful in erosion control of river banks.

49.5 CAJEPUT-TREE (*Melaleuca Leucodendra* L.)

Has thick spongy bark and leaves that taper at both ends.

49.5

THE EUGENIA GENUS (*Eugenia* L.)

Some of the 600 evergreen species of trees and shrubs in this genus bear edible fruits.

49.6 SURINAM-CHERRY; PITANGA
(*Eugenia uniflora* L.)

In this species the fruit is 8-ribbed and when ripe acquires a deep crimson color and a spicy taste.

49.6

Two other interesting trees in this family occur in the PIMENTA genus.

PIMENTO; ALLSPICE (*Pimenta dioica*)
The dried unripe berry is known as Allspice.

BAY- or BAY-RUM TREE (*Pimenta racemosa*)
The oil distilled from the leaves is used in the preparation of bay rum.

Although this is a rather large family of about 3,000 species in 175 genera, it is known in the U.S. only by virtue of one genus (*Rhexia*) that is native in the East, and by several other genera grown in hothouses. For the most part they are trees, shrubs, and herbs of tropical distribution.

THE GLORY-BUSH GENUS (*Tibouchina* Aubl.)

There are some 215 species of these, and their center of distribution is in Brazil. The leaves are large and leathery with 3 to 7 prominent veins. The flowers are large and showy.

50.1 GLORY-BUSH (*Tibouchina semidecandra* Cogn.)

The reddish-purple flowers are from 3 to 5 inches across and occur either singly or in groups of three at the ends of the branchlets. Beneath them are two orb-shaped bracts.

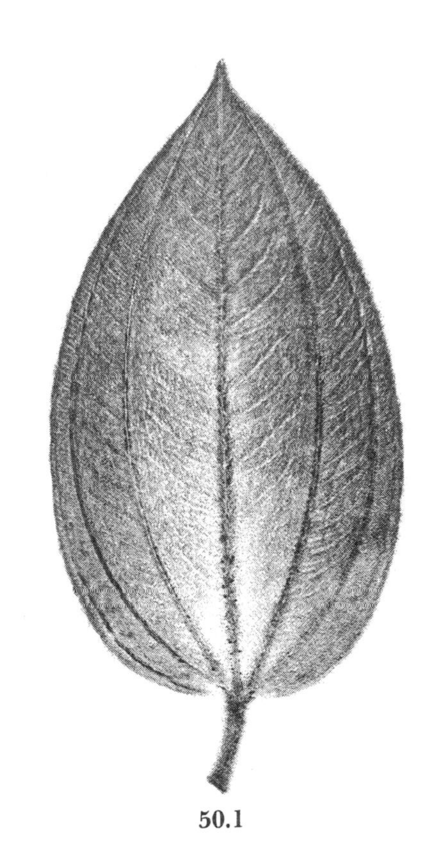

50.1

In the 10 genera and some 90 species of this family are to be found trees, shrubs, and some herbs mostly native in the northern hemisphere. The wood is very tough and the fruit a 2-seeded drupe or berry.

THE DOGWOOD GENUS (*Cornus* L.)

Forty species are included in the genus. Among the familiar ones, the leaves are bitter and the veins arched. Their fruits may be red, white, or blue (as well as green and black). Some authors warn against the danger of roasting hot dogs on a stick cut from **FLOWERING DOGWOOD** because of possible allergic reaction to a toxic substance in the wood. They are among the most popular ornamental trees in cultivation and, where they grow wild, are a glorious sight in springtime.

51.1a and 51.1b FLOWERING DOGWOOD (*Cornus florida* L.)

What appear to be the petals of the flower are in reality bracts (modified leaves). There are red, white, and pink-flowering varieties. The bark of small roots provides a red dye and the chewed twigs are said to make an excellent toothbrush (but note the caution above). Some 36 species of birds have been recorded feeding on its berries, which are also edible (although very bitter) for humans.

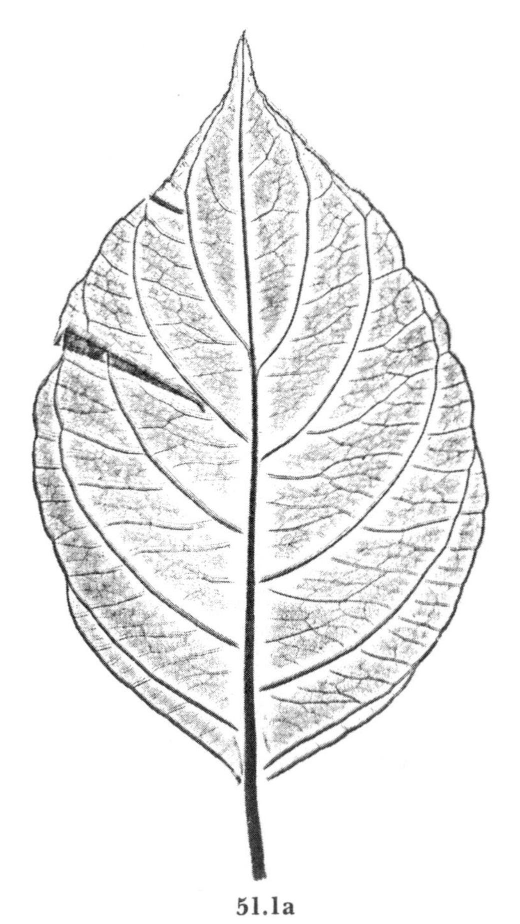

51.1a

51.1b

51.2 CHERRY CORNEL or CORNELIAN CHERRY (*Cornus mas* L.)

The flowers are yellow, the fruits scarlet and edible.

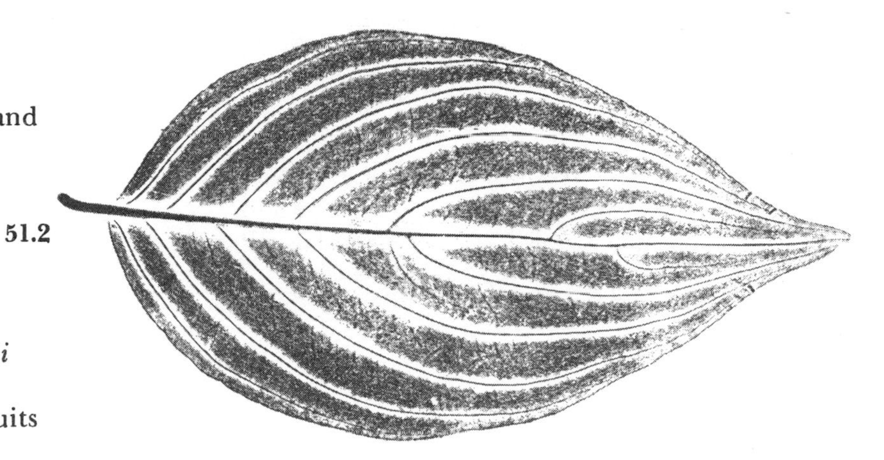
51.2

51.3 ROUGH DOGWOOD (*C. Drummondi* Meyer)

The leaves are sandpapery above, the fruits white, eaten by many songbirds.

51.3

51.4 KOUSA DOGWOOD (*C. Kousa* Hance.)

The flowers are creamy-yellow, the fruits pinkish. These are excellent for attracting birds.

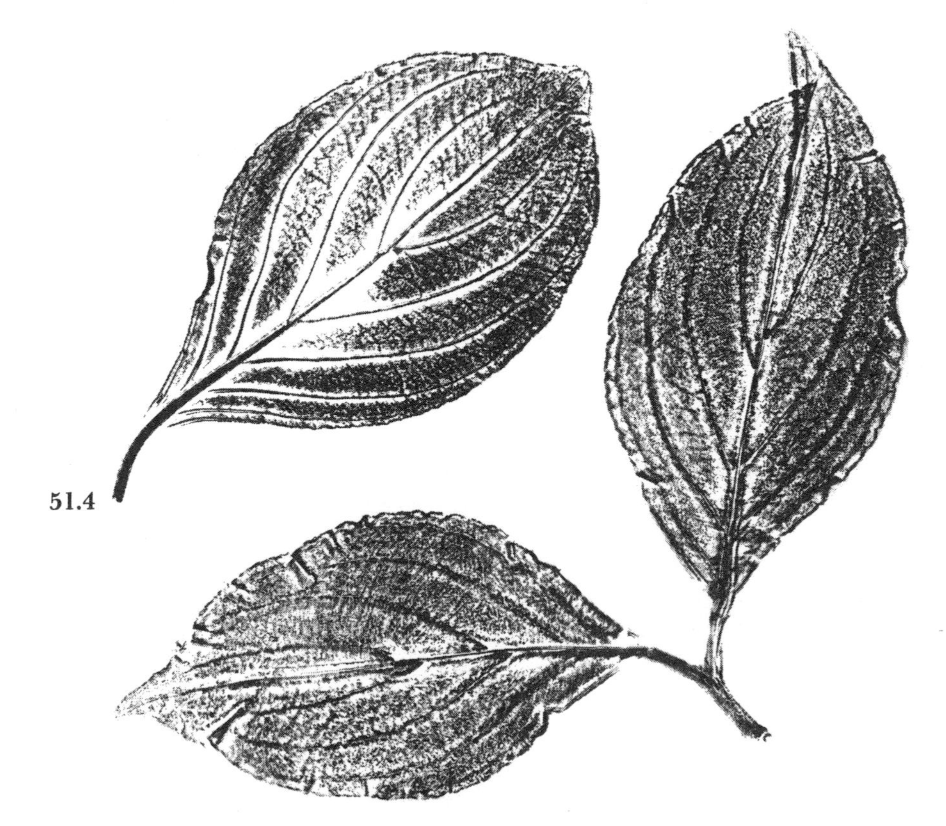
51.4

51.5 PAGODA DOGWOOD (*Cornus alternifolia* L.f.)

The flowers are creamy-white, the fruits dark blue.

51.5

THE AUCUBA GENUS (*Aucuba* Thunb.)

Opinion varies as to whether or not this genus consists of one very variable species. Bailey accepts a split into 3 separate species. Those in cultivation are usually tub-grown for their ornamental value. The small purple flowers grow in panicles; the fruit is a one-seeded, berry-like drupe.

51.6 GOLD-DUST TREE (*Aucuba japonica* Thunb. var. *variegata* D'Ombrain)

The leaves are spotted with yellow.

51.6

The 30 genera and about 550 species of Myrsines are widely distributed throughout the tropics and subtropics. Representatives of the genus described below have become popular as greenhouse plants.

THE ARDISIA GENUS (*Ardisia* Sw.)

These evergreen trees and shrubs are assigned to some 200 different species. They grow best in warm to hot climates.

52.1 ARDISIA (*Ardisia crenata* Sims)

The white or pink flowers occur on special lateral branches. The small fruits are coral-red. The leaves have crisped edges and for this reason the plant is sometimes confused with the species *Ardisia crispa* which, oddly enough, does *not* have crisped edges.

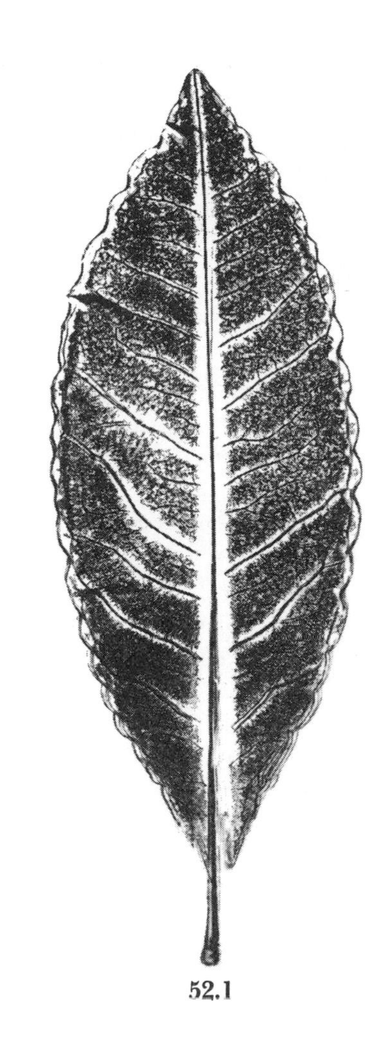

52.1

About 35 genera and 425 species of trees and shrubs constitute this widely distributed tropical family. Several are noted for their edible fruits; one (*Palaquium Gutta*) yields the nonelastic rubber called gutta-percha used in the construction of submarine cables, golf balls, and in a number of other indispensable ways.

53.1

THE ACHRAS GENUS (*Achras* L.)

A genus of evergreen trees widely cultivated for their edible fruits.

53.1 SAPODILLA (*Acras Zapota* L.)

This tall evergreen is a native of the Yucatan Peninsula. From its bark comes a latex that is about 25% gum, called chicle. It is the basis of the chewing-gum industry. In addition, the tree also produces a pear-like fruit with rusty-brown skin and yellow-brown flesh. It is considered one of the best fruits of the American tropics.

53.2 CANISTEL; EGG-FRUIT (*Lacuma nervosa* A.DC.)

Bears a 2- to 4-inch-long fruit with soft orange flesh and 2 or 3 shining seeds.

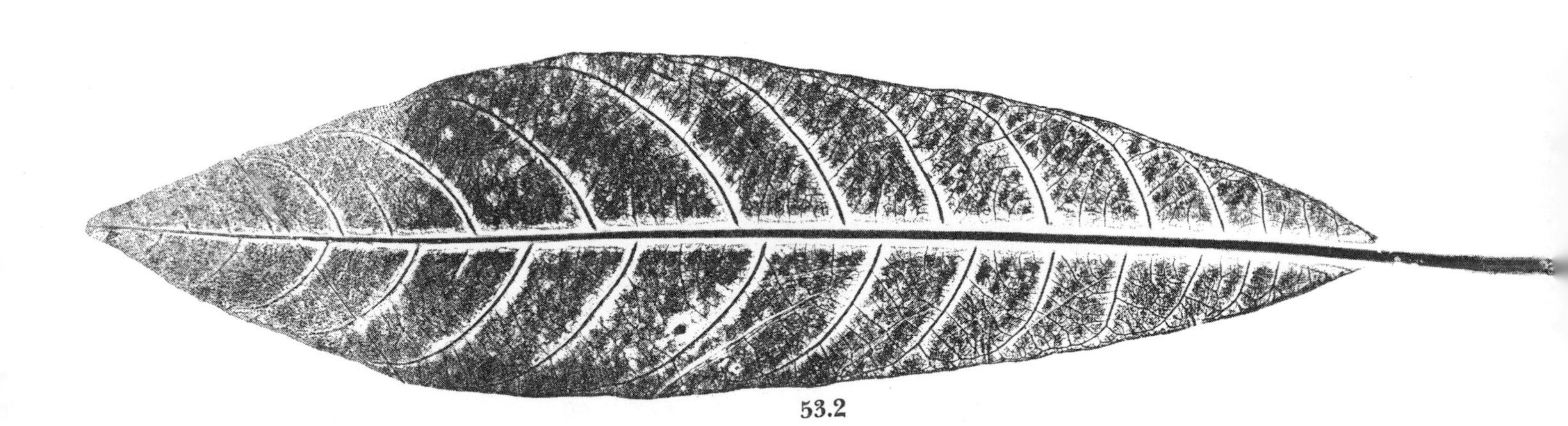

53.2

Of the 6 genera and about 300 species in this family of very hard-wooded trees and shrubs, one genus is noted for its edible fruit.

THE PERSIMMON GENUS (*Diospyros* L.)

The 200 or so species are widely distributed and are famous for their large, juicy, 1 to 10 seeded berries.

54.1 AMERICAN PERSIMMON (*Diospyros virginiana* L.)

An important American tree whose black heartwood is a true ebony. The extremely austere unripe fruits are used medicinally as an astringent. There is no point in the common statement that the fruits require frost before ripening. They ripen in due season, as do their tropical relatives which are never exposed to freezing. The ripe fruits may be preserved or mixed with flour and baked. The seeds can be roasted and used as a coffee substitute, according to Gillespie. If the outer coat of these seeds is scraped off, a pearly white core remains.

54.2 DATE-PLUM (*D. Lotus* L.)

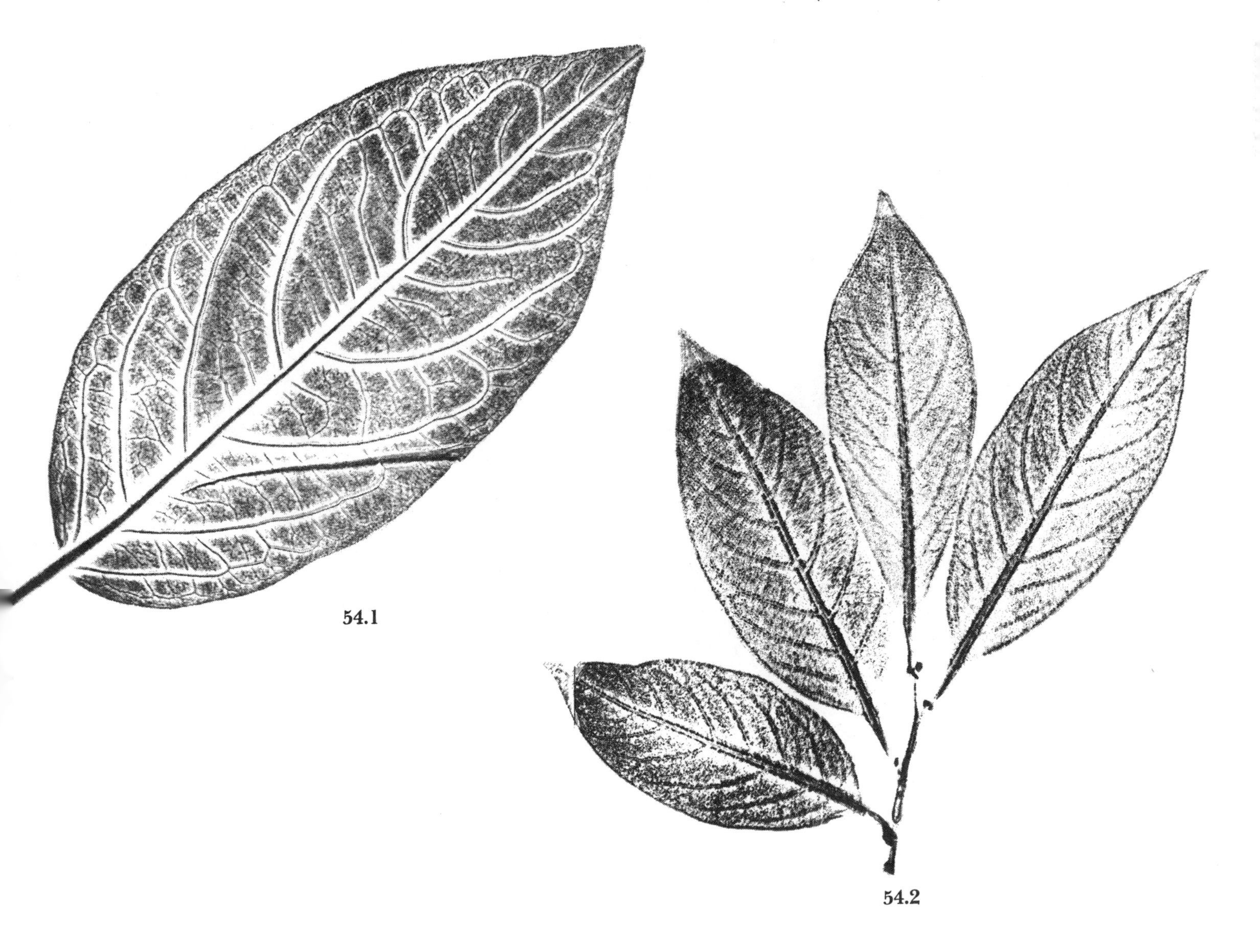

54.1

54.2

There are a few ornamental trees and shrubs of note among the members of the 6 genera and about 120 species in this family. In spite of the implication of the family name, most of the storax, or styrax, of commerce is not obtained from these trees but rather from the genus LIQUIDAMBAR in the family HAMAMELIDACEAE (Family 24 herein).

THE STORAX OR SNOWBELL GENUS (*Styrax* L.)

The 100 or so species are usually grown for their attractive flowers. The deciduous and evergreen trees and shrubs in the genus are native in tropical and subtropical regions.

55.1 AMERICAN STORAX (*Styrax americana* Lam.)

The white flowers bloom from April to June, from Virginia to Florida and Louisiana.

55.1

THE SILVER-BELL OR SNOWDROP-TREE GENUS (*Halesia* Ellis)

The 4 or 5 species in this genus are native in North America and China.

55.2 SILVER-BELL; SNOWDROP-TREE (*Halesia monticola* Sarg.)

A large tree that may reach 100 feet. Its leaves are 3 to 6 inches long. The white flowers bloom in April and May and the 2-inch-long fruits that follow are 4-winged.

55.2

This is a very important, though not very large, family of nearly 20 genera and 500 species that is found widely distributed in the temperate and tropical regions. The family is noteworthy particularly for the one genus that furnishes the edible olive, which has provided mankind with both oil and food since antiquity. Other species of olive are also useful in conservation work as bird foods; still others are important timber trees. A few are merely ornamental.

THE OLIVE GENUS (*Olea* L.)

About 40 species of evergreen trees of the tropics and warm regions of the Old World and New Zealand. Now grown everywhere throughout the Mediterranean region and grown in California since 1769, being commercially important since 1890. Olive trees live to a great age, but they require careful cultivation with a deep fertile soil and irrigation if necessary.

56.1 OLIVE (*Olea europaea* L.)

The olive of commerce.

56.1

THE FRINGE-TREE GENUS (*Chionanthus* L.)

Only two species, native in North America and China, are included in this genus.

56.2 FRINGE-TREE or WHITE-FRINGE (*Chionanthus virginica* L.)

An ornamental native bush-tree. The sexes are on separate specimens and the male or staminate flowers are showier. Females may produce blue berries that are not poisonous but are too bitter for humans to eat. They do have wildlife value as food for birds, which are the chief disseminators. It is quite important as a medicinal plant. Felter, as spokesman for the Eclectic School of medicine (which later became absorbed by "regular" medicine), is very strong in his praise of the drug made from the tree as a "specific medicine" for jaundice. In Appalachia the boiled root-bark liquor is used for skin irritations.

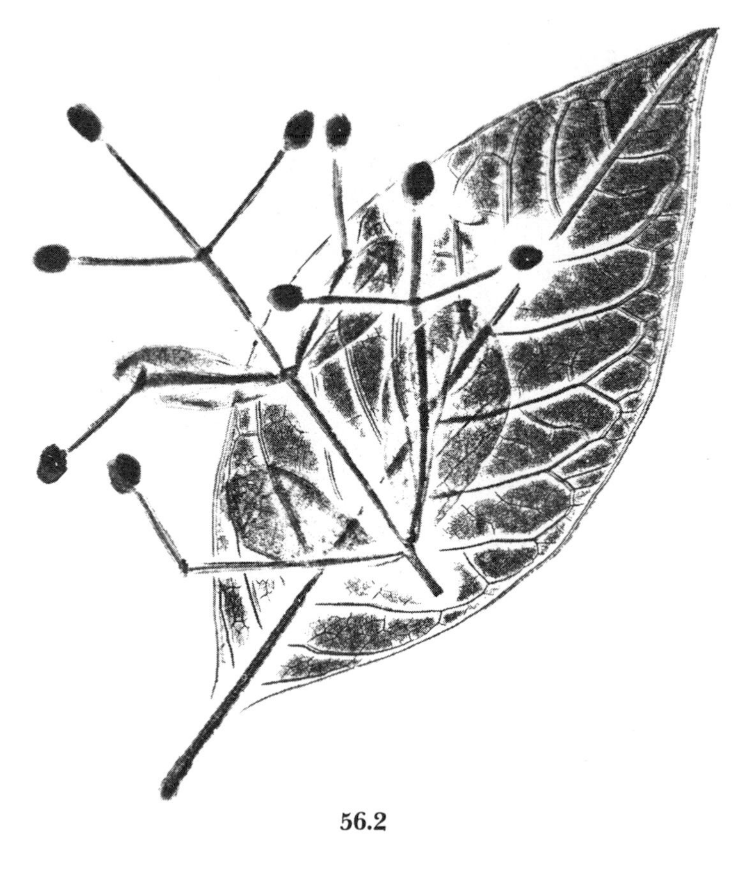

56.2

56.3 CALIFORNIA PRIVET (*Ligustrum ovalifolium* Hassk.)
This is widely planted as a hedge.

56.3

THE FORSYTHIA OR GOLDEN BELLS GENUS (*Forsythia* Vahl.)

About 8 species of deciduous shrubs that are widely planted for their masses of yellow early spring flowers, which bloom before the leaves appear. Very amenable to forcing.

56.4 FORSYTHIA (*Forsythia* sp.)

56.4

THE LILAC GENUS (*Syringa* L.)

The shrubs and small trees of the 30 species in this genus are justly popular for their large terminal clusters of flowers. The COMMON LILAC, shown here, has been favored for many decades. The site of an abandoned farmhouse can often be recognized by a lilac tree found growing in the woods. With the lilac as a clue, it usually does not take long to find the crumbled foundation stones and sometimes even the old well.

56.5 COMMON LILAC (*Syringa vulgaris* L.)

56.5

THE ASH GENUS (*Fraxinus* L.)

There are about 65 species of trees in this genus, which are popular for street and park planting. One species also has some use as a timber tree, since its wood is strong, elastic, tough, and light in weight. It is easy to split and hard to nail and so is not generally used for construction. Oars, bats, skis, tennis rackets, and baskets are acceptably produced from ash wood.

56.6 WHITE ASH (*Fraxinus americana* L.)

As a timber tree, the most important species in the genus. Some claim that rubbing a leaf on a mosquito bite eliminates the itch. Others use a poultice made from the inner bark on fever sores and a tea from the buds for snakebite. (Try the mosquito remedy first!)

BLACK ASH (*F. nigra* Marsh.)

Justly famous as the source of the wood used in making pack and other kinds of baskets. The original technique was to pound a log of BLACK ASH with a wooden mallet until it separated into layers corresponding to the annual rings. It is hard work. A refinement of the technique is to cut a board of desired length and width out of the log. The annual rings in this board should lie at right angles to its sides. As the pounding proceeds, every slat that comes off will be of the same width and length.

56.7 EUROPEAN ASH (*F. excelsior* L.)

Not to be confused with the EUROPEAN MOUNTAIN-ASH, which is in the Rose family.

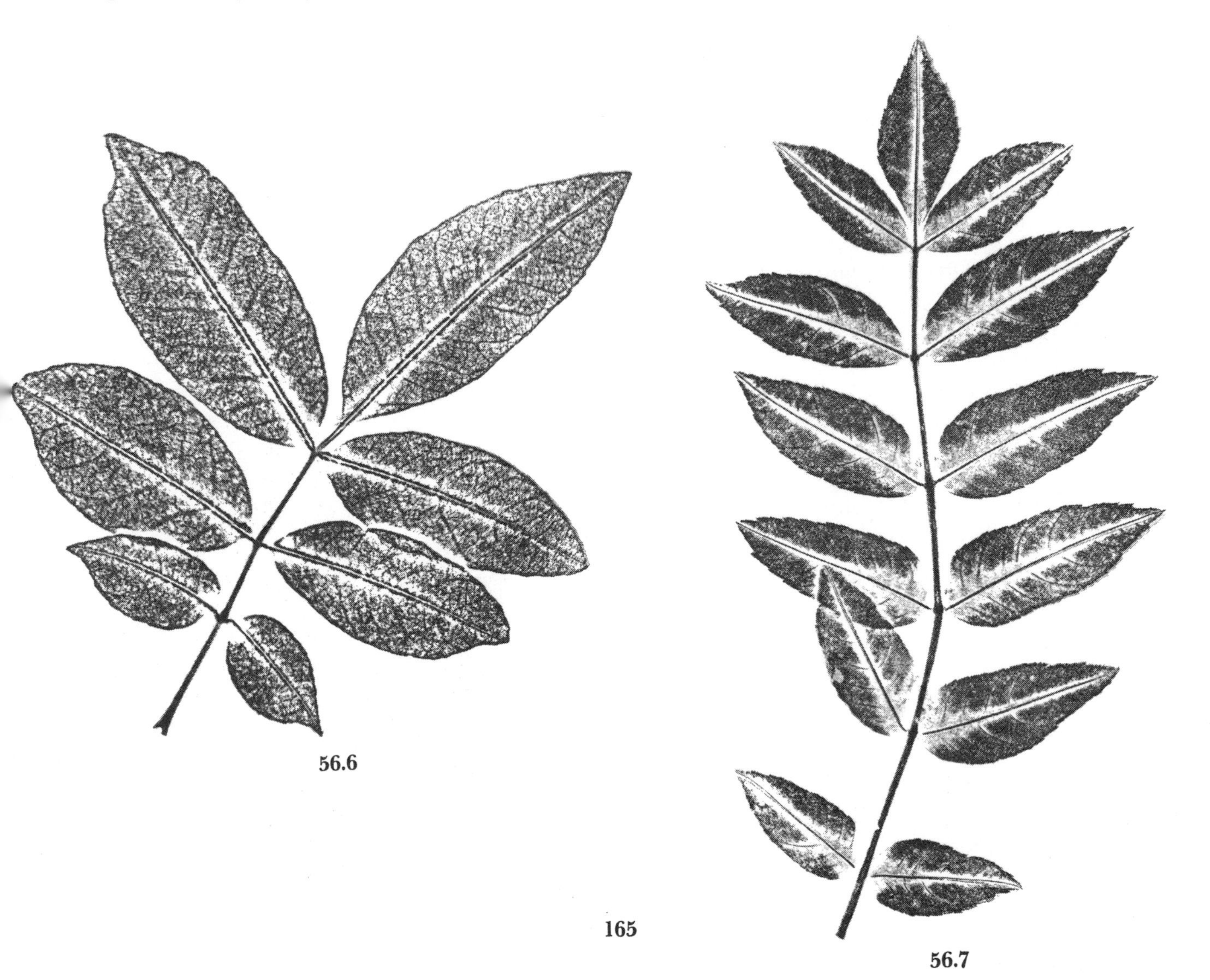

56.6

56.7

56.8 RED ASH (*Fraxinus pennsylvanica* Marsh.)

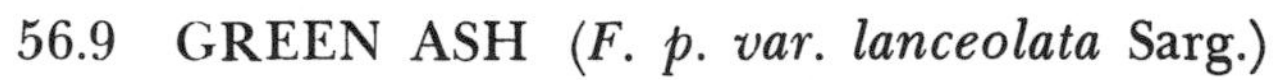

56.9 GREEN ASH (*F. p. var. lanceolata* Sarg.)

56.10 BLUE ASH (*F. quadrangulata* Michx.)

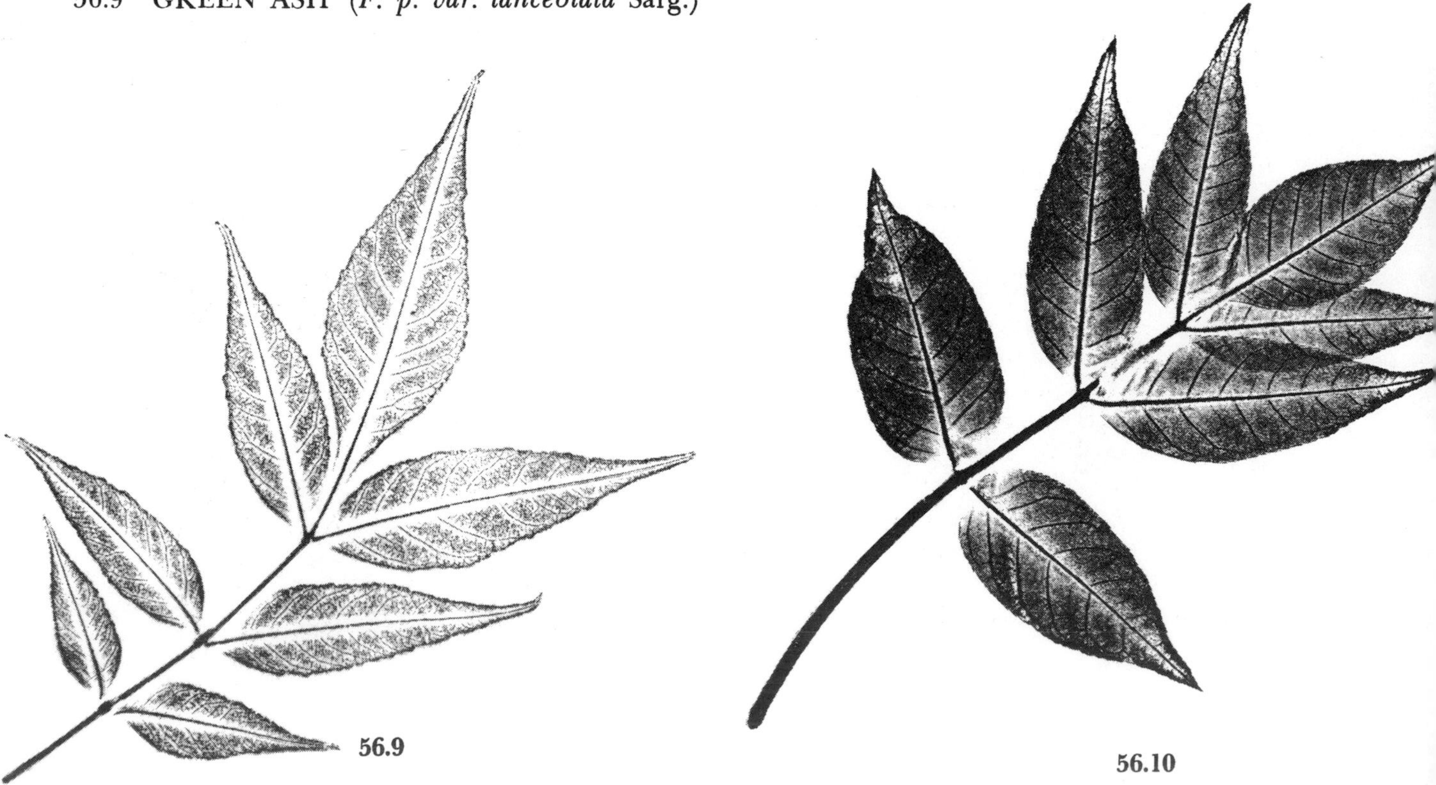

OLEACEAE

56.11 FLOWERING ASH (*Fraxinus Ornus* L.)

56.12 BILTMORE ASH (*F. americana* var. *biltmoreana* (Beadle) J. Wright)

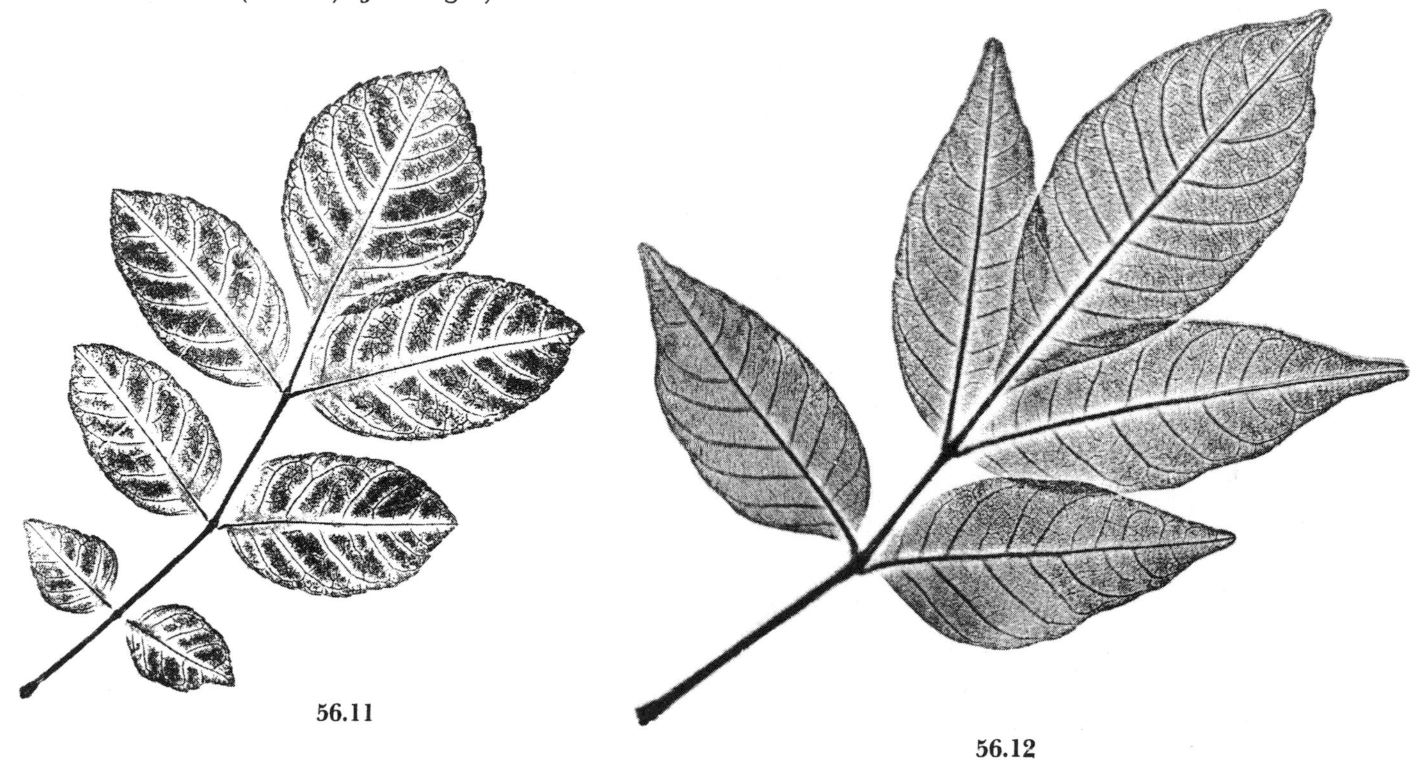

56.11

56.12

THE JASMINE GENUS (*Jasminum* L.)

There are about 200 species of tropical and subtropical deciduous shrubs in this genus. The species illustrated here has 4-angled, stiff branches and bears yellow flowers. These blossom out on the leafless shoots in late winter or early spring.

56.13 HARDY JASMINE (*Jasminum sp.* probably *J. nudiflorum*)

JASMINE (*J. officinale* L. var. *grandiflorum* Bailey)

This plant, extensively cultivated in France, is the source of the essential oil of Jasmine, which is removed from the flowers by a process called *enfleurage*. In this technique the flowers are left for several days lying on cold glass plates covered with fat, which dissolves and absorbs the perfume material from the petals. The essential oil is then recovered from the fat.

56.13

APOCYNACEAE

Herbs, trees, and shrubs are included in this family of about 300 genera and over 1,300 species. Some are climbers, some produce edible fruits, and some are grown as ornamentals. They are found widely scattered throughout the world, but are most abundant in tropical regions.

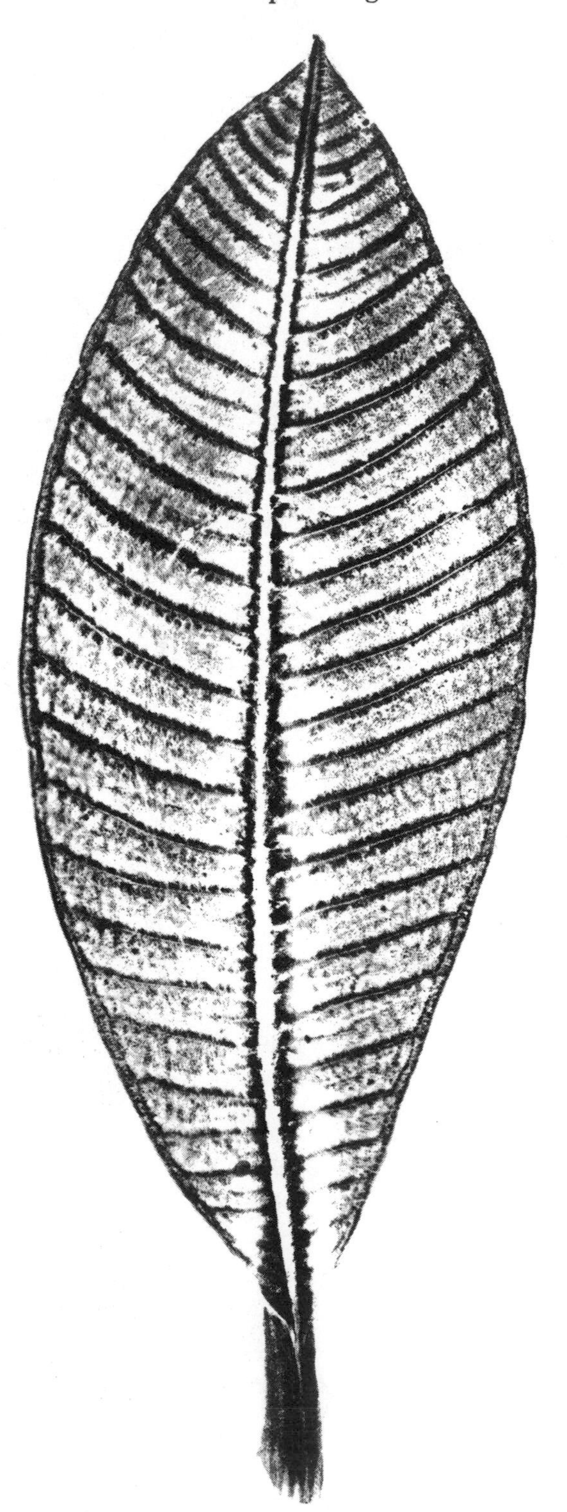

57.1

THE FRANGIPANI GENUS (*Plumeria* L.)

Consists of about 7 species of ornamentals from warm regions. These trees and shrubs have thick and fleshy branches with prominent leaf scars.

57.1 FRANGIPANI (*Plumeria rubra* L.)
Note the definite marginal vein.

THE ALLAMANDA GENUS (*Allamanda* L.)

About 12 species of climbing shrubs mostly native in Brazil but also grown in the Deep South in the U.S. or elsewhere under glass. They have large and showy flowers.

57.2 ALLAMANDA (*Allamanda* sp.)

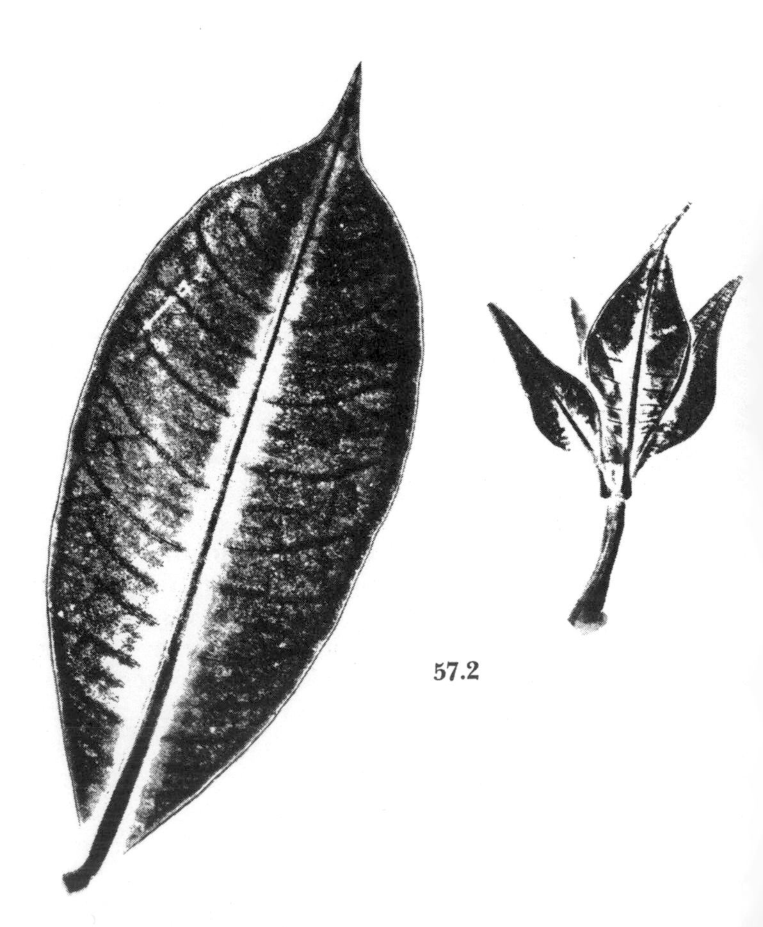

57.2

THE OLEANDER GENUS (*Nerium* L.)

Three species of upright small trees constitute the genus. Used as a house plant in the north, the leathery leaves are usually in whorls of three.

57.3 OLEANDER (*Nerium Oleander* L.)

All parts of this plant are extremely poisonous and eating a single leaf can have fatal results. The wood should not be used to roast meat over open fires, et cetera. Even honey made from the nectar of the flowers is poisonous.

THE CARISSA GENUS (*Carissa* L.)

Some 30 species of spreading shrubs, a few of which are used in the South as hedge plants; others may reach 18 feet in height. As hedge plants, the species shown here is especially useful because of its 2-pronged 1½-inch-long spines.

57.4 NATAL-PLUM (*Carissa grandiflora* A.DC.)

The scarlet 1- to 2-inch-long berries are edible and have been used for the making of jellies and preserves.

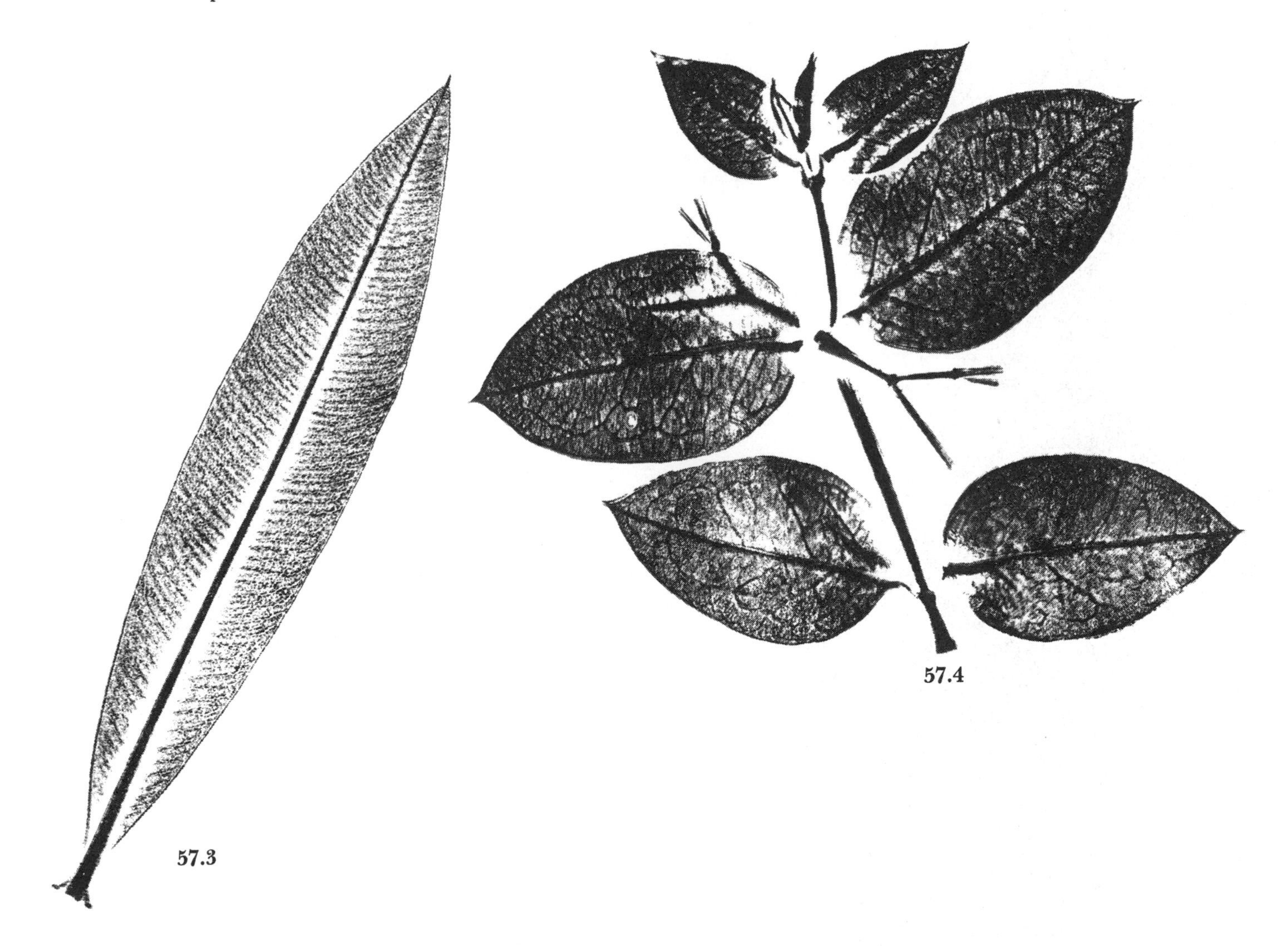

57.3

57.4

THE AMSONIA GENUS (*Amsonia* Walt.)

This genus consists of 17 species of perennial herbs. The species shown is sometimes transplanted into gardens for its blue, many-flowered cymes.

57.5 BLUE AMSONIA (*Amsonia Tabernaemontana* Walt.)

THE OCHROSIA GENUS (*Ochrosia* Juss.)

One species of this very small genus is seen in Florida as a cultivated tree, although its native home is in Australia and the Pacific Islands. It has leathery leaves. A violet-like odor emanates from the scarlet fruits when they are crushed.

57.6 OCHROSIA (*Ochrosia elliptica* Labill.)
This is sometimes also called the OCHROSIA PLUM. The striking red-skinned and white-fleshed drupes are considered to be poisonous.

57.5

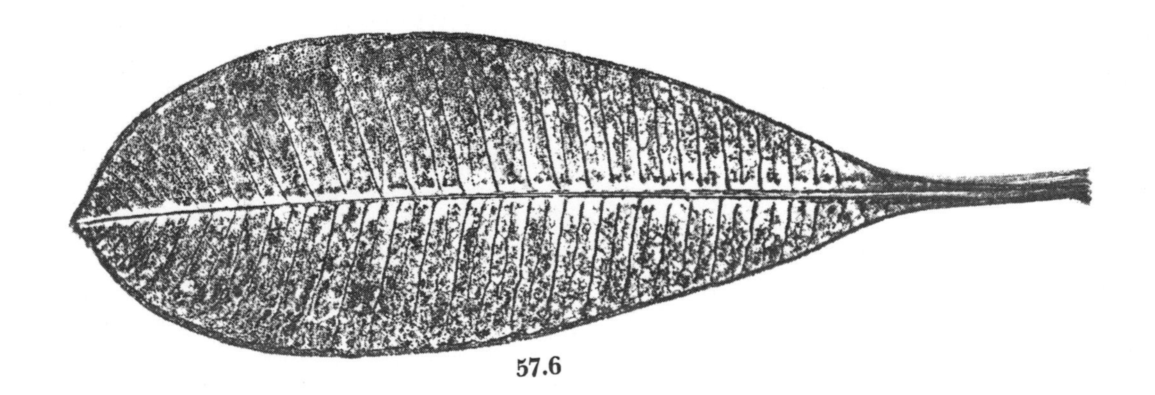

57.6

Trees, shrubs, and vines are included in this family of about 100 genera and 600 species, but only a few are native in the temperate zone. The flowers are irregular in construction and quite showy in effect. The fruits are long, bean-like capsules whose seeds are usually winged.

THE CRESCENTIA GENUS (*Crescentia* L.)

There are about 5 species in the genus native in tropical America, but also seen cultivated in the southern U.S.

58.1 CALABASH-TREE (*Crescentia cujete* L.)
The fruit of this tree has been used for water-gourds since antiquity. The fruit is a hard-rinded berry containing many wingless seeds within a pulpy mass. This material is removed to fashion a water-gourd.

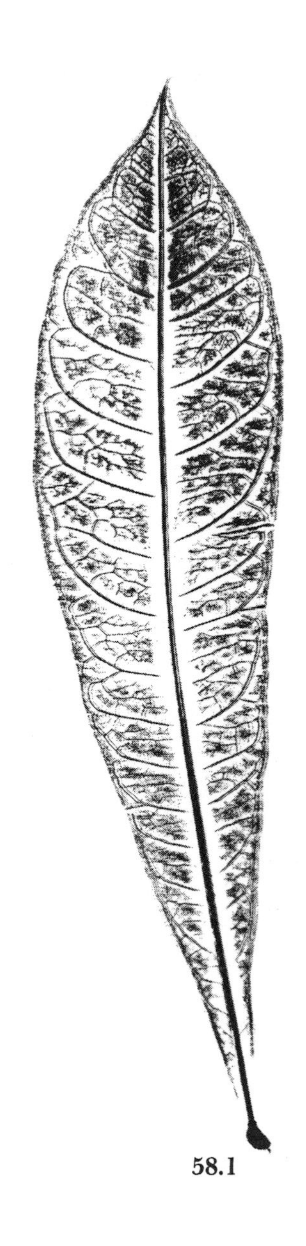

58.1

THE CATALPA GENUS (*Catalpa* Scop.)

Some of the 10 species of this genus may be seen in timber plantations. The others are used ornamentally.

58.2 YELLOW CATALPA (*Catalpa ovata* Don)
The flowers are striped with orange inside and spotted with dark purple. They are widely planted in Japan.

58.2

58.3 COMMON CATALPA; INDIAN BEAN (*C. bignonioides* Walt.)

The flowers are white, with two yellow stripes and many purplish-brown spots.

58.4 TALL CATALPA (*C. speciosa* Warder.)

The flowers are larger than above, the spots not very apparent. The petal lobes are frilled at the margins.

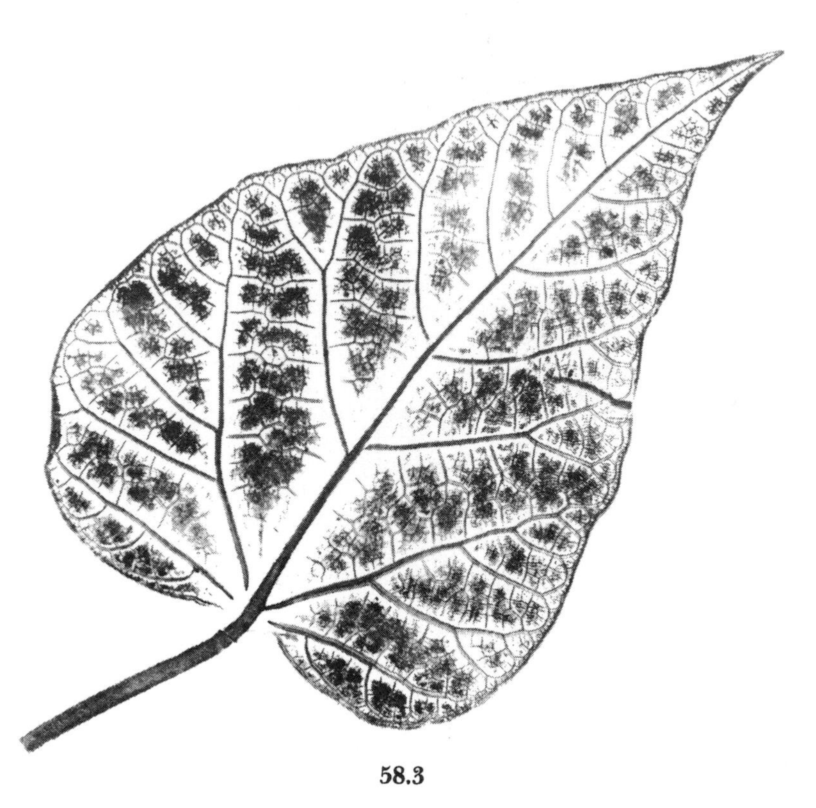

58.3

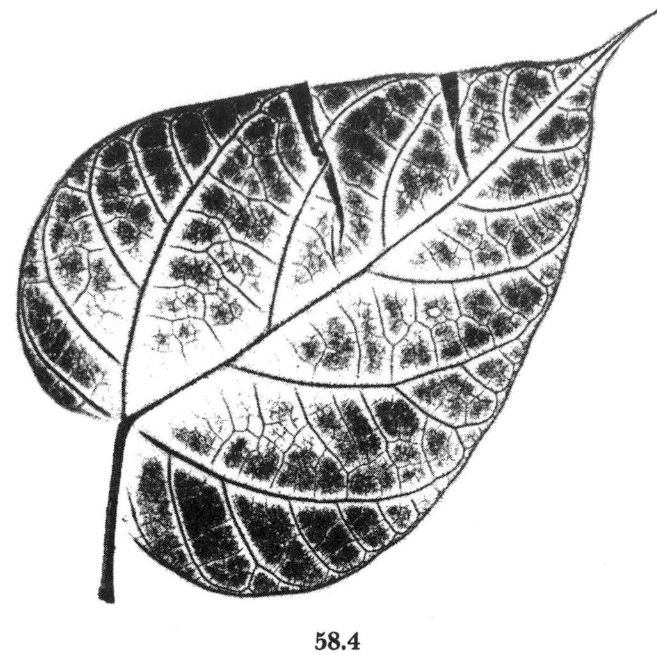

58.4

This is a large family containing some 4,500 (Benson) to 5,000 (Bailey) species. It includes about 425 genera, some of which are solely ornamental. A number of the species produce bitter medicines, and coffee and quinine from this family are commercial products of inestimable importance. There are both woody and herbaceous species in the family and many (but not all) of the woody species are tropical or subtropical. This is an instance of the general principle that if species of a plant family occur in both tropical and temperate zones, the former will be chiefly woody and the latter will be largely herbaceous.

THE COFFEE GENUS (*Coffea* L.)

A rather small group of 25 to 40 species of evergreen trees and shrubs whose taxonomic relationships have been confused by horticultural practices.

59.1 COFFEE (*Coffea arabica* L.)

A small bushy tree native in northern Africa but now widely cultivated in Brazil and other American countries. The story of coffee is fascinatingly told by Thornton. The derivative chemical caffeine is present in the cola drinks and in many headache relief remedies.

59.1

THE BUTTON-BUSH GENUS (*Cephalanthus* L.)

Only one species is cultivated out of the 6 species of these ornamental woody shrubs.

59.2 BUTTONBUSH (*Cephalanthus occidentalis* L.)

A native shrub, also cultivated for ornament. The flowers are fragrant and a source of honey. The bark is bitter, used as a bitter tonic and febrifuge. Leaves turn red and other colors. The seed clusters are used for decorative purposes.

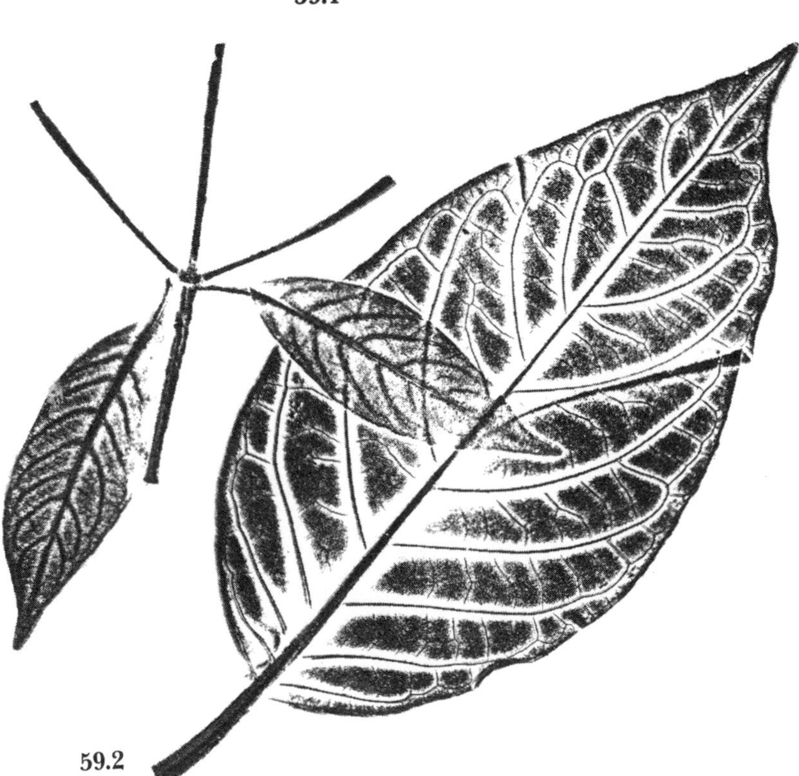

59.2

THE IXORA GENUS (*Ixora* L.)

Over 150 species of these evergreen shrubs and small trees are distributed in the tropical regions of the world.

59.3 FLAME-OF-THE-WOODS (*Ixora coccinea* L.)

This species is one of the commonest landscape shrubs around Miami, where it is used especially for hedges and as a landscape specimen. The leaves are in pairs but hard to print that way. The flowers are brilliant red and very showy. Other species of *Ixora* with larger flowers (red, pink, or yellow) are also in cultviation. The blackish berries are dry but edible. They are reputed to be relished by monkeys.

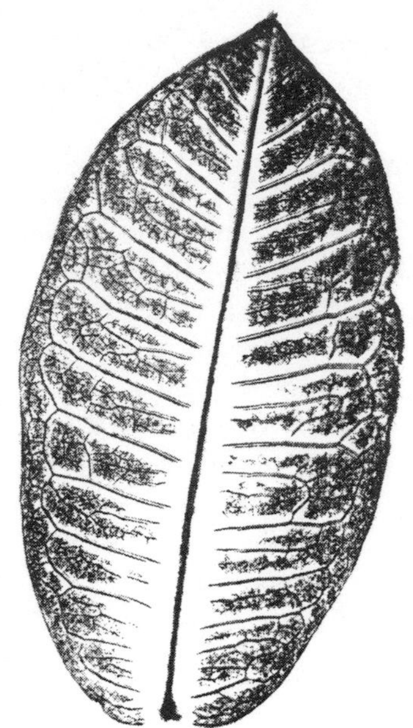

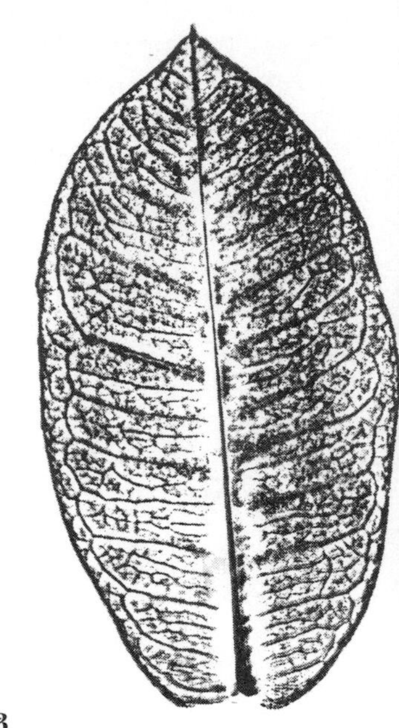

59.3

There are about 400 species in some 14 genera in this family of shrubs and a few herbaceous plants. For the most part they are ornamentals, but a few have been used medicinally.

THE ELDERBERRY GENUS (*Sambucus* L.)

There are approximately 20 species in this genus.

60.1 SWEET ELDER: ELDERBERRY (*Sambucus canadensis* L.)

This is potentially an important shrub, for it yields immense quantities of safely edible berries which can be made into preserves and beverages. All other parts produce enough hydrocyanic acid to be considered dangerous. The flowers are harmless and have been fried as a delicacy. The species has a hollow stem that has been used in various ways by children. Poisoning may result if the stems are placed in the mouth. (See also Tehon, page 101.) The fruits give a purple dye. They are second only to raspberries and blackberries in their attractiveness to birds.

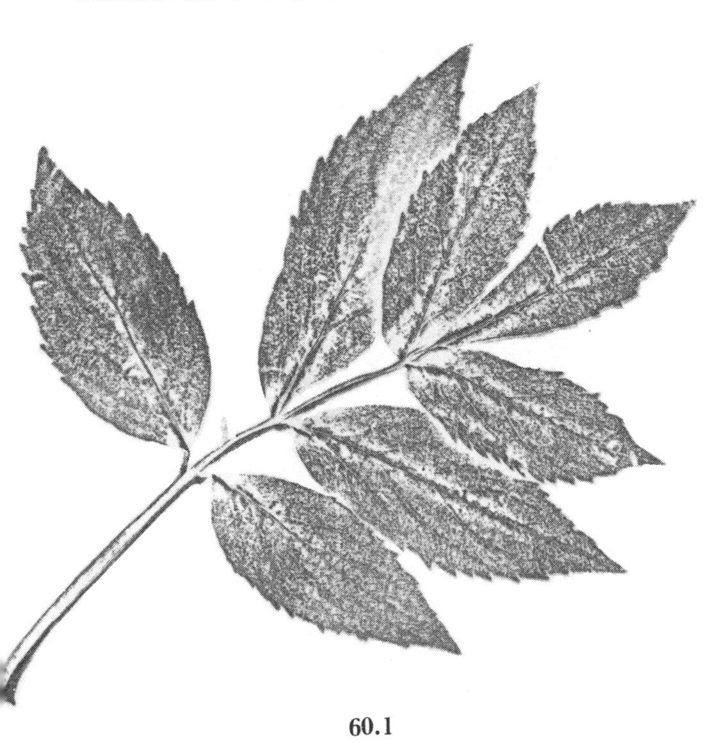

60.1

60.2 CORALBERRY or INDIAN-CURRANT (*Symphoricarpos orbiculatus* Moench)

A very important native shrub for conservation purposes. The spreading and suckering manner of growth make it a good landhold for erosion control, the denseness affords shelter for wildlife, and the berries are a food for birds. The reddish berries are, however, too bitter, dry, and pulpy for humans, but they are not poisonous. This shrub is a weed in pastures but is cultivated for ornament.

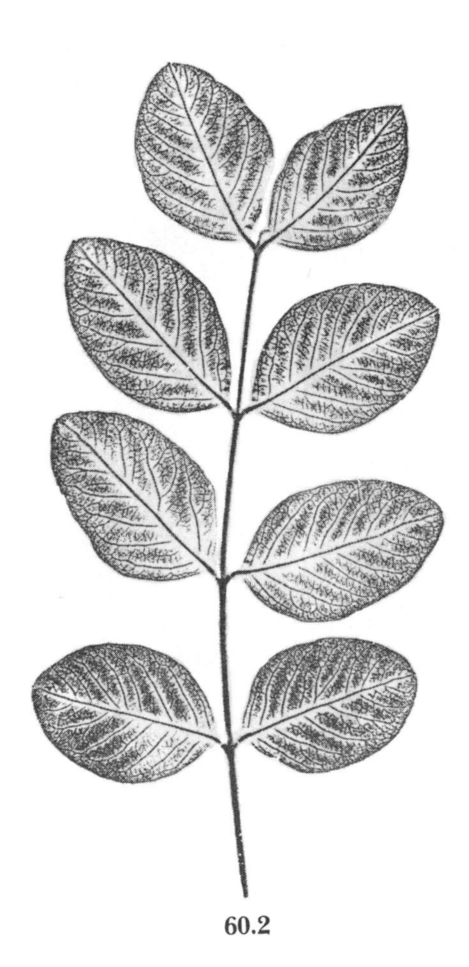

60.2

THE HONEYSUCKLE GENUS
(*Lonicera* L.)

In this genus there are about 180 species of shrubs that vary from deciduous to evergreen, and from climbing to erect.

60.3 JAPANESE CLIMBING HONEYSUCKLE
(*Lonicera japonica* Thunb.)
Essentially a weed, although the fruits are eaten by many birds and animals that also enjoy its dense cover.

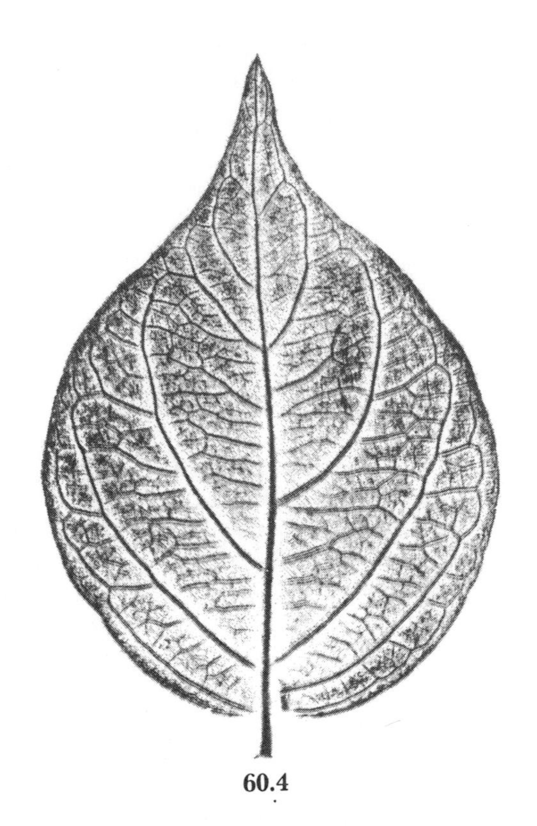

60.4

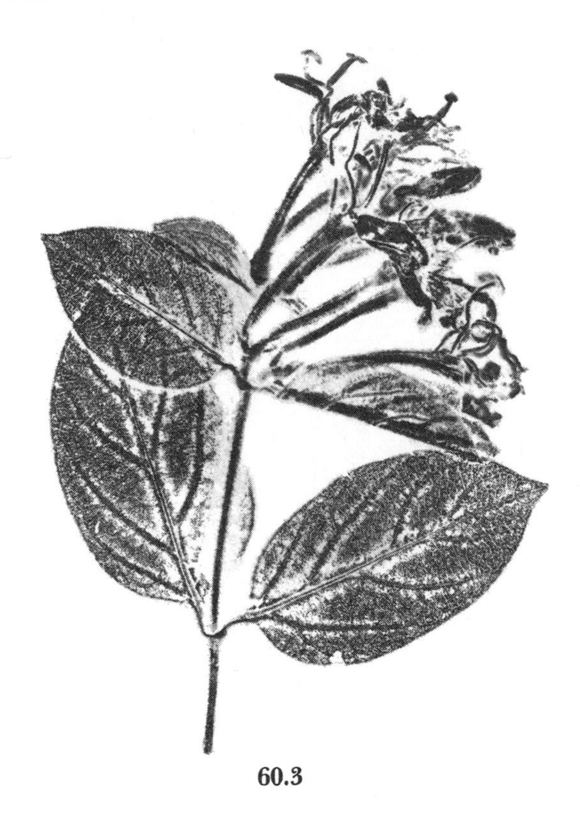

60.3

60.4 MAACK'S BUSH HONEYSUCKLE
(*L. Maackii* Maxim.)
Birds eat the fruits of this, also.

60.5 FRAGRANT BUSH HONEYSUCKLE
(*L. fragrantissima* Lind. & Paxt.)

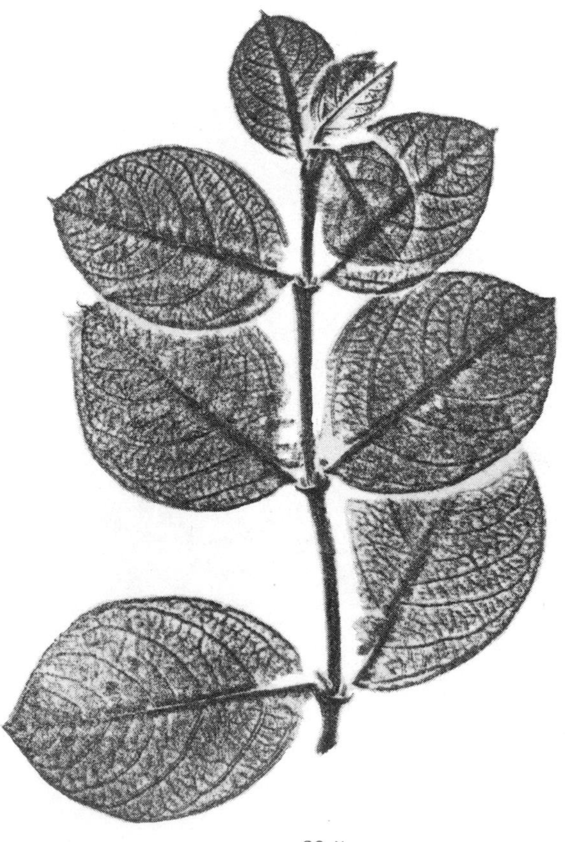

60.5

60.6

60.7

60.7 WITHE ROD (*Viburnum cassinoides*)
Has blue-black berries.

60.8 HOBBLEBUSH (*V. alnifolium,* Marsh.)
Has dark purple berries.

60.6 BLUE-BUSH HONEYSUCKLE
(*L. caerulea* L.)

THE VIBURNUM GENUS
(*Viburnum* L.)

About 120 species of attractive ornamental shrubs. The fruit is a one-seeded drupe that is produced in a variety of colors by the various species. Also known by the names HOBBLEBUSH, NANNYBERRY, HAW, WILD RAISIN, and SQUASHBERRY, these shrubs provide edible fruits for eating raw or in jellies and pies.

60.8

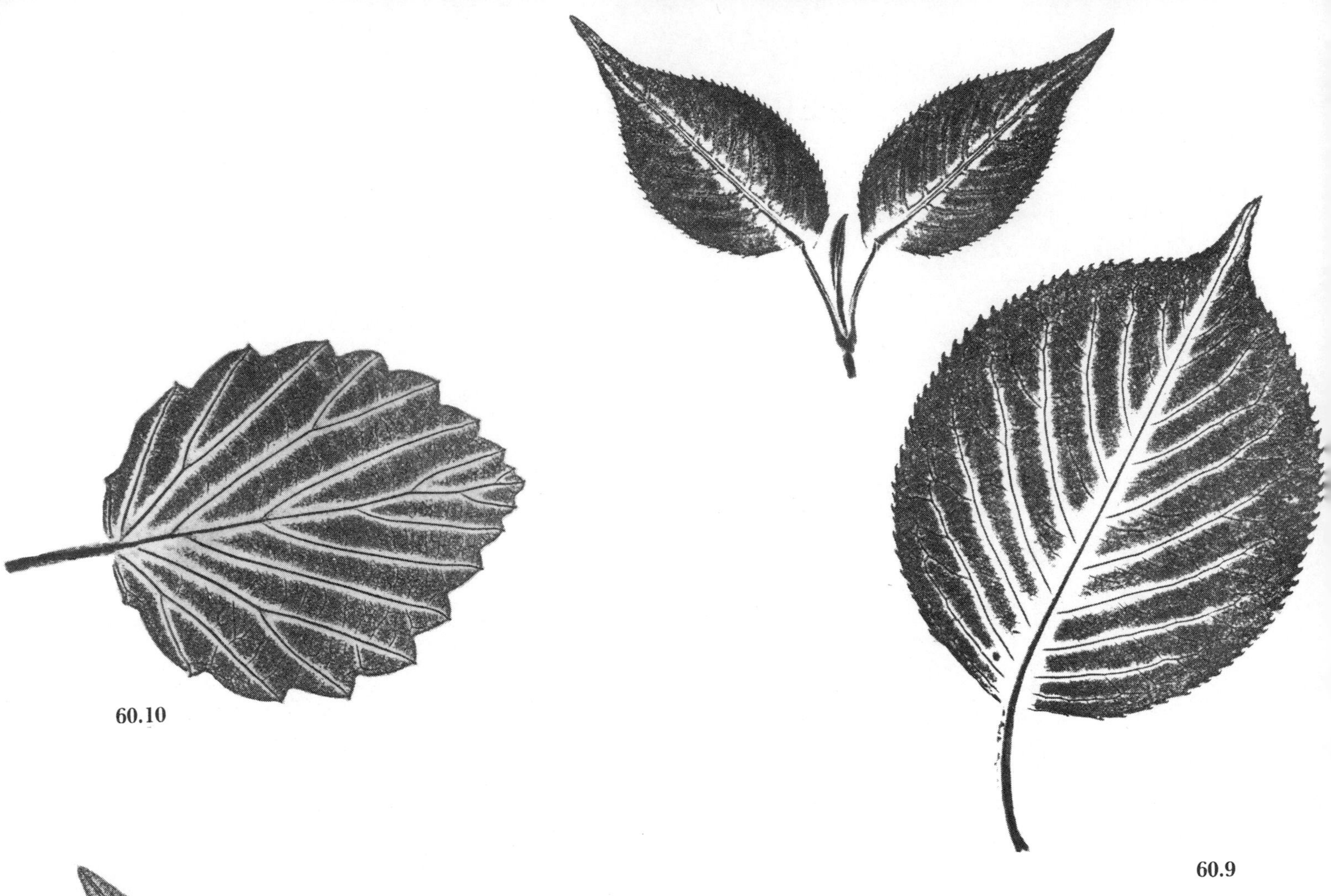

60.10

60.9

60.11

60.9 SWEET VIBURNUM; SHEEP-BERRY; NANNY-BERRY (*Viburnum Lentago* L.)

60.10 SOUTHERN ARROW-WOOD (*V. dentatum* L).

Bluish-black berries. Both of these species are used to attract the rose-breasted grosbeak.

60.11 BUSH-CRANBERRY (*Viburnum trilobum* Marsh.)

The HIGHBUSH-CRANBERRY, CRANBERRY-BUSH, or *Viburnum americanum* (of some nurserymen) bears bright scarlet berries that keep their color until the following spring. They are a favorite food of the cedar waxwing and are sometimes used as a substitute for true cranberries. The bark is medicinal.

60.12 SIEBOLD'S or CUCUMBER VIBURNUM (*V. Sieboldii* Miq.)

60.13 WAYFARING-TREE (*Viburnum Lantana* L.)

Bears bright red berries, which gradually change almost to black.

60.12

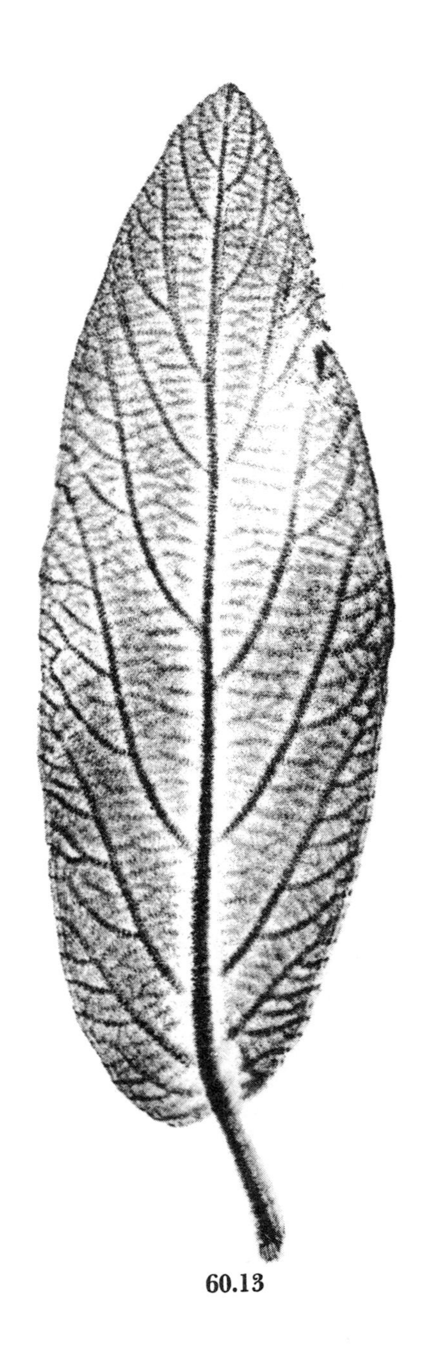

60.13

Some Technical Terms Defined

Abscission. The separation of leaves or fruits from a branch as a result of the growth of a specific layer of cells—the abscission layer.

Aggregate fruit. As in blackberry, a fruit formed from numerous pistils that were separate in the flower.

Albumen. Synonym for the endosperm (food-storing) region of the seed.

Aril. Fleshy outgrowth of the hilum or funiculus that partly or entirely surrounds a seed.

Berry. A pulpy fruit developed from one pistil, with one or more seeds.

Bract. A modified (often scale-like) leaf subtending a flower.

Carpel. One of the modified leaves from which a pistil is formed (if compound). A simple pistil has but one carpel.

Circinate. An unrolling type of plant growth seen in the "fiddle heads" of ferns.

Compound (leaves). A leaf having more than one blade attached to its petiole.

Cordate. Heart-shaped.

Cyme. A rather flat-topped flower cluster whose central flowers open first.

Deciduous. A tree that sheds all of its leaves at one season so that the tree is bare for a portion of the year.

Dioecious. When the two sexes are housed in separate (plant) bodies.

Distichous. Two-ranked in the same plane.

Drupe. A stone-fruit. The fleshy indehiscent fruit contains one seed enclosed in a stony endocarp.

Drupelet. As in the raspberry, one of the drupes of the aggregation that makes up the fruit.

Dehiscent. Applied to anthers or seed pods that open naturally along specific lines.

Embryo. The developing plantlet within the seed.

Endocarp. The inner layer of the ovary wall.

Evergreen. Trees whose leaves do not fall all at one time, so that they always appear to be fully-leaved.

Fascicle. A bundle or cluster.

Funiculus. A stalk that attaches a seed to the ovary wall.

Herbaceous (plants). Nonwoody plants, usually annuals.

Hilum. A scar on a seed marking the point of attachment of the funiculus.

Lenticels. Corky breathing-pores of a stem.

Marcescent. Leaves that wither but remain attached to the branch.

Meiosis. A type of cell division that reduces the chromosome count in the nucleus by one-half.

Monoecious. Having the organs of the two sexes on the same (plant) body.

Ovary. The basal, enlarged portion of a pistil that contains the ovules.

Ovule. A megasporangium. After the egg of the megagametophyte that grows within the megasporangium is fertilized, a seed develops.

Megagametophyte. A tiny egg-producing plant that grows within a megasporangium.

Megasporangium. The structure within the ovary that produces a spore that develops into a megagametophyte.

Palmate (leaf). A leaf whose major veins meet at a common point at the base of the leaf.

Pinnate. Constructed on the same plan as a feather.

Petiole. The stem-like portion of a leaf. May be long or short.

Pistil. The usually centrally located (so-called female) part of a flower whose basal ovary contains one or more ovules.

Pyrene. The nutlet (seed) in a drupe.

Raceme. An elongated inflorescence of stalked flowers in which the topmost flower opens last.

Rachis. The central stem-like axis of a pinnately compound leaf.

Samara. Indehiscent winged fruit most easily recognized as the so-called nose of the Maple genus.

Scale (leaves). Small dry vestigial leaves closely applied to the stem.

Seed. A ripened ovule consisting of the embryo and its coats, which are the persistent megasporangial walls.

Serrate. Saw-toothed.

Spores. Cells that develop into plants without the need of fertilization.

Stipule. Usually two basal appendages on the petiole of a leaf. May be modified into spines or may be absent.

Syncarp. A fruit formed by the union of several carpels.

Truncate. As though almost squarely cut off.

Vermiculate. As though tunneled through by worms.

Some Books for Further Study

Only a small part of the story of the usefulness of trees and shrubs is told in this herbal. It is but an introduction to a fascinating segment of human knowledge. Many of the books listed here will introduce you to strange plants growing in strange places. Start with this herbal and then go on to the other books; your life may never be quite the same again.

The quarterly magazine *Economic Botany* is published by the New York Botanical Garden for the Society for Economic Botany, 3110 Elm St., Baltimore, Md. Separate addresses are given for membership, subscriptions, and back issues—all obtainable from the one given above. This is a most valuable publication covering and extending the topics in this herbal. It would unquestionably be the best place for anyone interested in the general subject of economic botany or any of its phases to begin his research.

The commercial weekly *Oil, Paint and Drug Reporter* is published by the Auchincloss family under the business name Schnell Publishing Company, New York City 10007. To the student of economic botany, its most important feature is a long alphabetical list in every issue covering the asking and selling prices of hundreds of botanical materials.

The standard old herbals are not included in this bibliography because not generally accessible to the public. For a comprehensive treatment of these see Agnes Arber's *Herbals, Their Origin and Evolution,* listed below. The Lloyd Library of Botany, Pharmacy and Materia Medica, 309 W. Court St., Cincinnati, Ohio 45202, has a superb collection of herbals and other pertinent books assembled in the bookstores of Europe by Curtis Gates Lloyd.

THE FOUR FOUNDATION STONES

Bailey, L. H. *Manual of Cultivated Plants.* New York: The Macmillan Co., 1949.

Benson, Lyman. *Plant Classification.* Boston: D. C. Heath and Co., 1957.

Fernald, Merritt L. *Gray's Manual of Botany.* New York: American Book Co., 1950.

Hill, Albert F. *Economic Botany.* New York: McGraw-Hill Book Co., 1952.

OTHERS FOR FURTHER STUDY

Arber, Agnes R. *Herbals, Their Origin and Evolution.* Cambridge: University Press, 1953.

Arena, Jay M. *Poisoning: Chemistry, Symptoms, Treatments.* Springfield, Ill.: Charles C. Thomas, 1963.

Betts, H. S. *American Woods.* Leaflets on individual tree species. United States Forest Service, Government Printing Office, 1945–1954.

Clute, Willard N. *Useful Plants of the World.* Indianapolis, Ind.: Clute, 1943.

Coon, Nelson. *Using Plants for Healing.* New York: Hearthside Press, 1963.

Edlin, H. L. *Woodland Crafts in Britain.* London: Batsford, 1949.

Fairchild, David. *The World Was My Garden.* New York: Scribner, 1938.

Felter, Harvey W. *The Eclectic Materia Medica, Pharmacology and Therapeutics.* Cincinnati, Ohio: Scudder, 1922.

Fitzpatrick, Frederick L. *Our Plant Resources.* New York: Holt, Rinehart & Winston, 1964.

Gillespie, William F. *A Compilation of the Edible Wild Plants of West Virginia.* New York: Scholar's Library, 1959. (Very valuable little book because Mr. Gillespie "compiled" more from people than from other books.)

Goodrich, Francis L. *Mountain Homespun.* New Haven, Conn.: Yale University Press, 1931.

Graves, Arthur H. *Illustrated Guide to Trees and Shrubs.* Published by the author. Wallingford, Conn., 1952. Rev. ed., Harper, 1956.

Hardin, James W., and Arena, Jay M. *Human Poisoning from Native and Cultivated Plants.* Durham, N.C.: Duke University Press, 1969.

Harlow, William M. *Trees of the Eastern United States and Canada.* New York: McGraw-Hill Book Co., 1942.

Harlow, William M., and Harrar, Ellwood S. *Textbook of Dendrology.* New York: McGraw-Hill Book Co., 1950; 4th ed., 1958.

Hough, Franklin B. *The Elements of Forestry.* Cincinnati, Ohio: Clarke, 1882. (Spoken of in this herbal as "the elder Hough.")

Hough, Romeyn B. *Handbook of the Trees of the Northern States and Canada.* New York: The Macmillan Company, 1947 ("the younger Hough.")

Hughes, G. Bernard. *Living Crafts.* London: The Lutterworth Press, 1953.

Jaeger, Ellsworth. *Nature Crafts.* New York: The Macmillan Co., 1950.

Johnson, Laurence. *A Manual of the Medical Botany of North America.* New York: W. Wood & Co., 1884.

Kephart, Horace. *Camping and Woodcraft.* New York: The Macmillan Co., 1949.

Krochmal, Arnold *et al.* *A Guide to Medicinal Plants of Appalachia.* Washington, D.C.: Gov't. Printing Office, 1971.

Leopold, Aldo. *A Sand County Almanac.* New York: Oxford University Press, 1966.

Lyons, A. B. *Plant Names, Scientific and Popular.* Detroit, Mich.: Nelson, Baker & Co., 1900. (Despite the name, contains authentic drug properties and uses of plants.)

Marderosian, Ara Der. "Poisonous Plants in and around the Home." *American Journal of Pharmaceutical Education* 30 (1966) : 115–40.

Martin, George W., and Scott, Robert W. *Food in the Wilderness.* Bremerton, Wash.: Martin, 1963.

Marx, David S. *The American Book of the Woods.* Cincinnati, Ohio: Botanic, 1940.

Masefield, G. B. *et al.* *The Oxford Book of Food Plants.* London: Oxford University Press, 1969.

Mason, Bernard S. *Woodcraft.* New York: A. S. Barnes and Company, 1939.

Medsger, Oliver Perry. *Edible Wild Plants.* New York: The Macmillan Co., 1939.

Merrill, Elmer D. *Plant Life of the Pacific World.* New York: The Macmillan Co., 1945.

Meyer, Joseph E. *The Herbalist.* Hammond, Ind.: Indiana Botanic Gardens, 1960.

Morton, Julia F. "Ornamental Plants with Poisonous Properties." *Proceedings of the Florida State Horticultural Society* 71 (1958) : 372–80; 75 (1962) : 484–91.

———. "Principal Wild Food Plants of the United States." *Economic Botany* 17 no. 4, 1963.

Oswald, Fred. W. *The Ranger's Guide to Useful Plants of the Eastern Wilds.* Boston: Christopher Publishing House, 1964.

Peattie, Donald C., ed. *A Natural History of Trees of Eastern and Central North America.* Boston: Houghton, Mifflin Co., 1950.

Rafinesque, C. S. *Medical Flora.* Philadelphia: Atkinson & Alexander, 1828–1830.

Rehder, Alfred. *Manual of Cultivated Trees and Shrubs Hardy in North America.* New York: The Macmillan Co., 1951.

Sargent, Charles S. *Manual of the Trees of North America.* Boston: Houghton, Mifflin Co., 1933.

Schoonover, Shelly E. *American Woods.* Santa Monica, Cal.: Watlin, 1951.

Shankin, Margaret E. *Use of Nature Craft Materials.* Peoria, Ill.: Manual Arts Press, 1947.

Sweet, Muriel. *Common Edible and Useful Plants of the West.* Healdsburg, Cal.: Naturegraph Co., 1962.

Tehon, Leo R. *The Drug Plants of Illinois.* Urbana, Illinois State Circular 44, 1951. ("Only a pamphlet," but reasonably complete and very usable. The best available handbook on the subject of wild drug plants.)

Thornton, Robert John. *A New Family Herbal.* London: Richard Phillips, 1810.

Van Dersal, William R. *The American Land—Its History and Its Uses.* New York: Oxford University Press, 1943.

Van Rensselaer, Eleanor. *Decorating with Pods and Cones.* Princeton, N.J.: D. Van Nostrand Co., 1957.

Wyman, Donald. *Trees for American Gardens.* New York: The Macmillan Co., 1951.

Yearbook Committee. U.S. Dep't of Agriculture. *Trees, the Yearbook of Agriculture, 1949.* Washington, D.C.: Gov't Printing Office.

Yoakley, Ina C. "Wild Plant Industry of the Southern Appalachians," *Economic Geography* 8: 358, 1932.

Youngken, Heber W. *Pharmaceutical Botany.* Philadelphia: Blakiston, 1951.

INDEX OF PLANT USES

Italic type is used for the Latin names of the plants.

INDEX OF PLANTS

Italic type is used for the Latin names of the plants.
Roman type is used for the common names of the plants.
General topics found in the Introduction or in the body of the text are printed in small capitals.